Kites

to make and fly

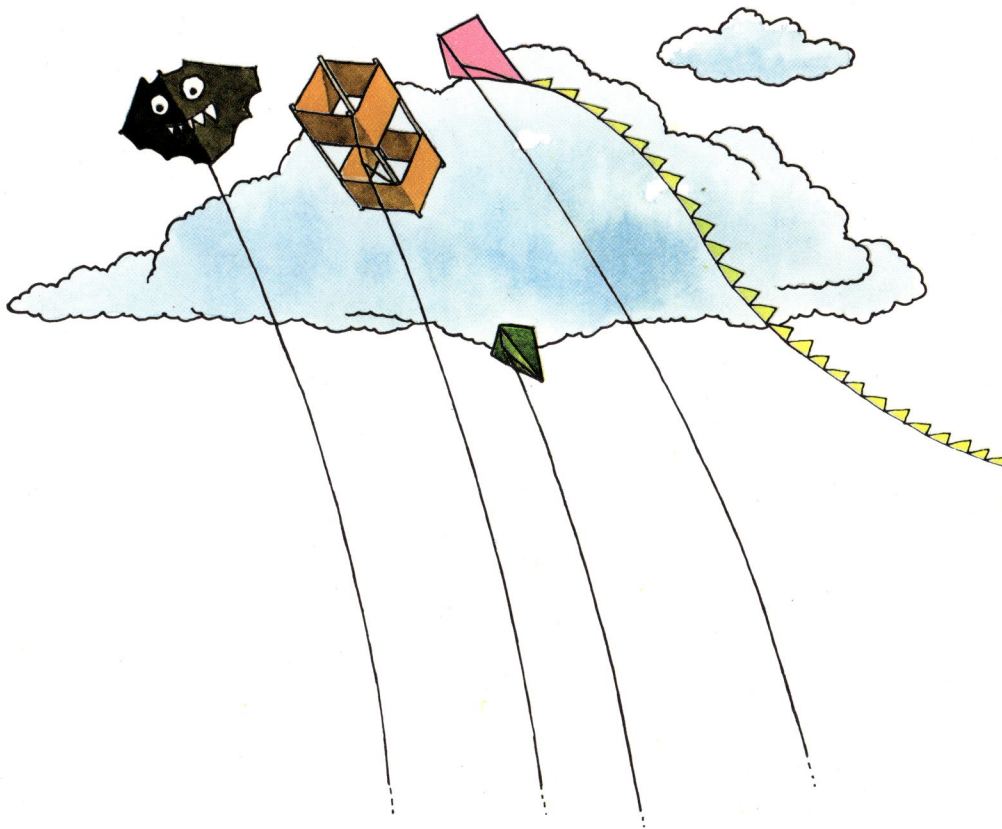

A Puffin Book
Written, illustrated and produced by Jack Newnham
Copyright © Penguin Books Australia Ltd, 1977

So you want to fly a kite

This book will show you how to make and fly four sorts of kite:

a flat kite with a tail
a bowed kite without a tail
and a box kite

There is also an unusual light-weight kite made with only one stick. It is called the **Minuteman** because it can be made quickly. Being very light, it only needs a gentle breeze to fly well. It is not strong enough to fly in a strong wind, but its strength may surprise you.

The **Tadpole** flat kite has a tail. In very strong winds it needs a very long tail. Without its tail, it just can't face the wind and goes into a dizzy spin.

There is one important thing to know about kites – it doesn't matter how good they look on the ground, a kite is not a kite until it is up in the air flying.

Some kites don't seem to **want** to fly. They are a bit like wild animals who need some sort of training to get them to perform.

Of course you can buy a ready-made kite. But you are not a real 'kitemaster' until you have made your very own and trained it to fly yourself.

FLY YOU RASCAL!

EEK!

THE 'TADPOLE' FLAT KITE

HI BIG BOY!

A bowed kite such as the **Bat** doesn't need a tail because its 'wings' curve back. This keeps it on course. Because it doesn't have the weight and 'drag' of a tail, it will fly almost directly over your head.

THE 'BAT' BOWED KITE

When it flies the Tadpole looks like a fish, and the Bat looks like a bat. But box kites are strange. Whoever heard of a flying box?

They might be strange – but box kites fly very well. In strong winds they seem to just 'hang' in the sky. The first successful aeroplanes were built after experiments with box kites.

Our box kite – we call it the **Sock Box** – is a little more complicated than the other kites. You might need an older person to help you build it.

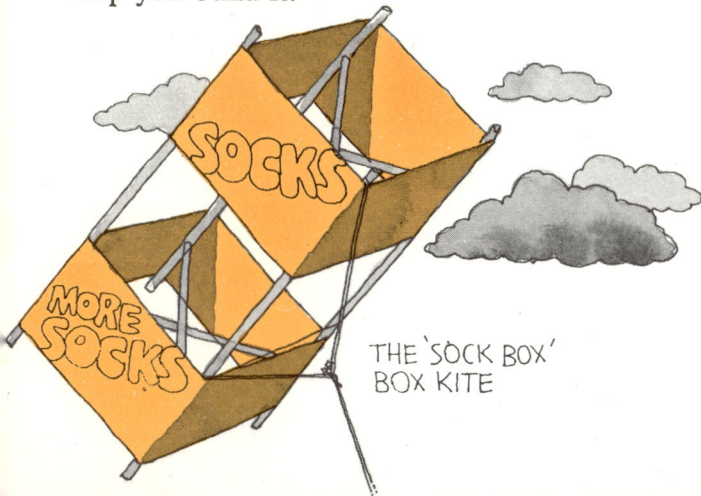

THE 'SOCK BOX' BOX KITE

What kites need to fly

All kites need **wind**.

They also need lots of **space**. If there are trees and buildings too near, they are sure to crash into them. Also the wind direction will keep changing making the kite confused.

Just as a ship's sails need setting to catch the wind, a kite's 'rigging' needs **adjusting** so that it will face the wind properly.

How each kite flies is discussed after the building instructions.

DON'T BLAME THE KITE IF THERE IS NO WIND

STUPID DARN KITE!

Safety

Kite-flying is great fun and quite safe, but here are some sensible rules to remember.

Always fly your kite in open space (park or countryside) well away from power lines.

Don't fly it when there is lightning about.

Never use wire for a flying line.

Don't let anyone stand near a wild-flying kite. Even a very light kite can hurt you.

Don't let the flying line run quickly through the fingers. The friction can burn.

Kites can be a danger to aircraft. Kite-flying is mostly forbidden within 3 kilometres of an airfield. In some areas you can't fly above a height of 70 or 80 metres.

The Tadpole Kite

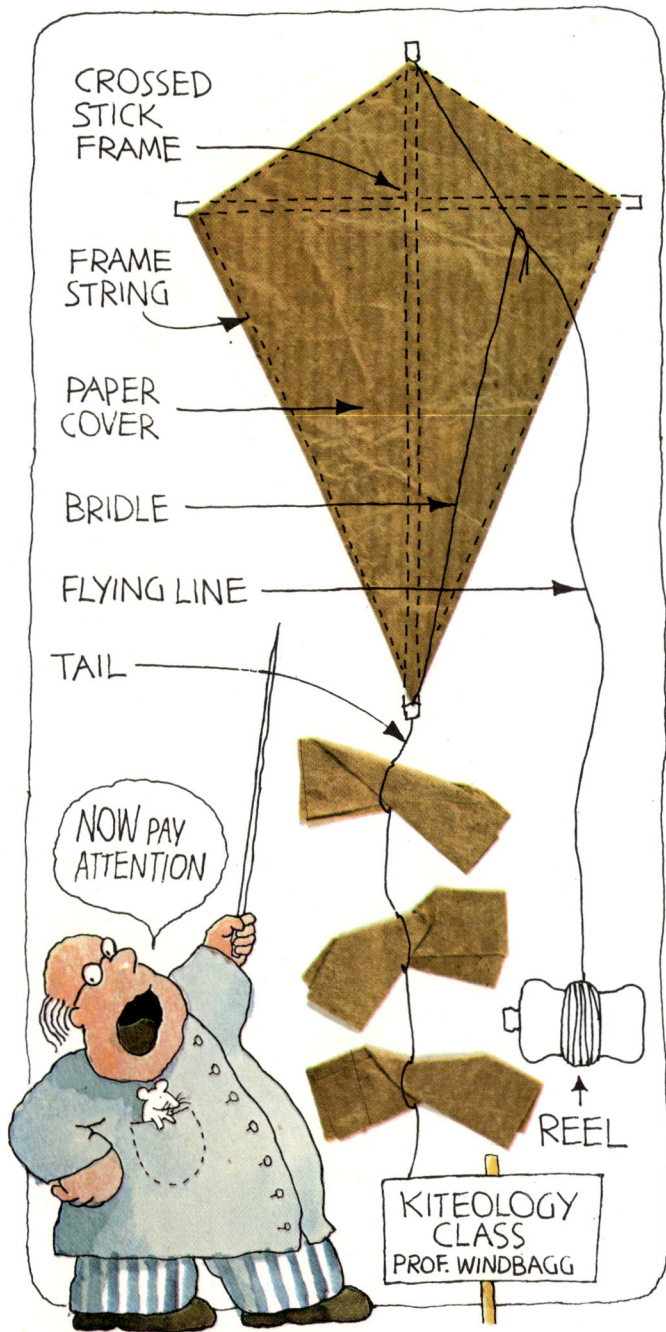

CROSSED STICK FRAME

FRAME STRING

PAPER COVER

BRIDLE

FLYING LINE

TAIL

NOW PAY ATTENTION

REEL

KITEOLOGY CLASS
PROF. WINDBAGG

Tools and things you will need

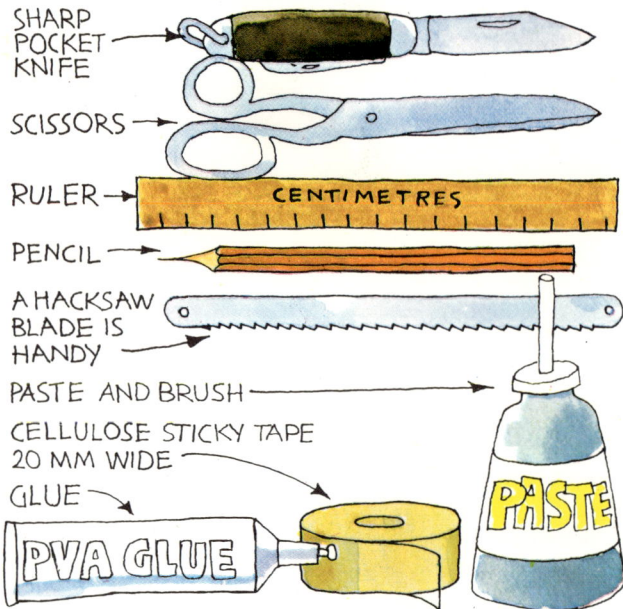

SHARP POCKET KNIFE

SCISSORS

RULER → CENTIMETRES

PENCIL

A HACKSAW BLADE IS HANDY

PASTE AND BRUSH

CELLULOSE STICKY TAPE 20 MM WIDE

GLUE

PVA GLUE

PASTE

The Tadpole and all our other kites are made of sticks, paper and string, with a few other odds and ends.

The paper is for the cover. Wrapping paper (white, brown or coloured) is ideal. It doesn't matter if it is creased. Small pieces can be overlapped and pasted together to make up the area if necessary. Holes can be patched and tears repaired with sticky tape. If you can't get any wrapping paper, a good Tadpole can be made with ordinary newspaper.

String. You will need about 100 metres of strong household string. Soft white cotton twine is better than the thick brown kind.

Make a **flying line reel**. Tie and wind your line to an empty plastic bottle or a wide piece of wood.

LEMON LIQUID

The sticks are used for the frame. They should be strong, straight and light. You should be able to bend them a bit without breaking them.

The upright stick is called the **spine**, the cross stick is the **spar**.

SPAR
SPINE

THE FRAME STRING
GOES AROUND THE OUTSIDE

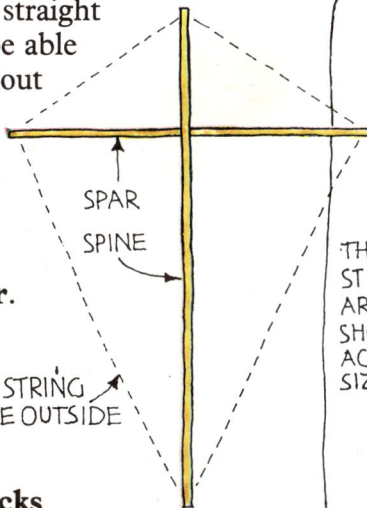

Where to get your sticks

Bamboo plant stakes about 1 metre long, 6–8 mm thick can often be bought from garden supply shops. One stick carefully split down the centre could make the 2 sticks required for the Tadpole, but buy 2 in case the split goes wrong and you can only use the thick piece.

Ready-made sticks: Timber moulding strips, bought from hardware stores make very good sticks. They must be flat on one side. Half-round moulding (called fly bead) 12 mm wide and 6 mm thick has been used for all the kites in this book, except the Minuteman. For the Tadpole buy a piece 1.5 metres long.

Hand-made sticks: Boards about 5 mm thick (and at least 72 cm long) from wooden packing cases can be split, if the grain is straight, to make kite sticks. Split off sticks from 1 to 2 cm wide and trim them straight with your knife to about 1 cm wide.

DON'T HAVE YOUR STICKS TOO THICK AND HEAVY THEY SHOULD BE ABOUT THE SAME THICKNESS AS SHOWN IN THIS DRAWING

END VIEW

6MM

5MM

8MM

12MM

1 CM

THESE STICKS ARE SHOWN ACTUAL SIZE

SPLIT BAMBOO STICK

FLY BEAD MOULDING READY-MADE STICK

HAND-MADE STICK SPLIT FROM BOARD

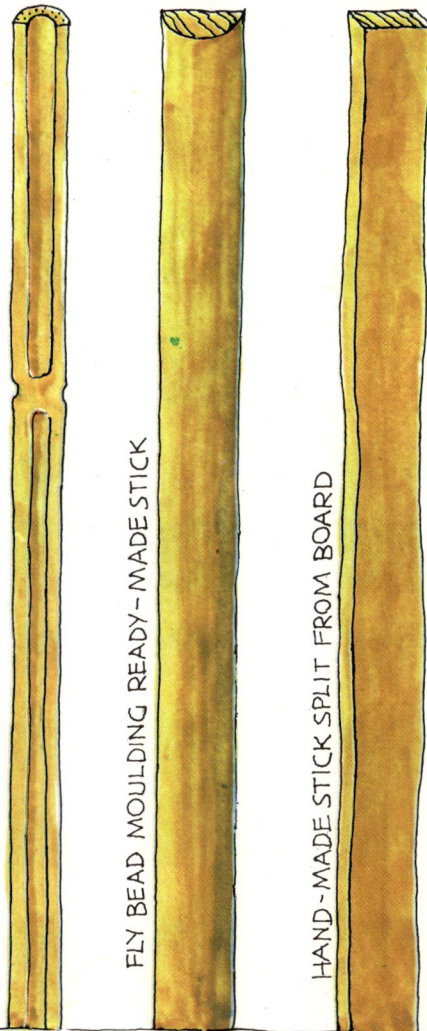

Cutting the sticks for the frame

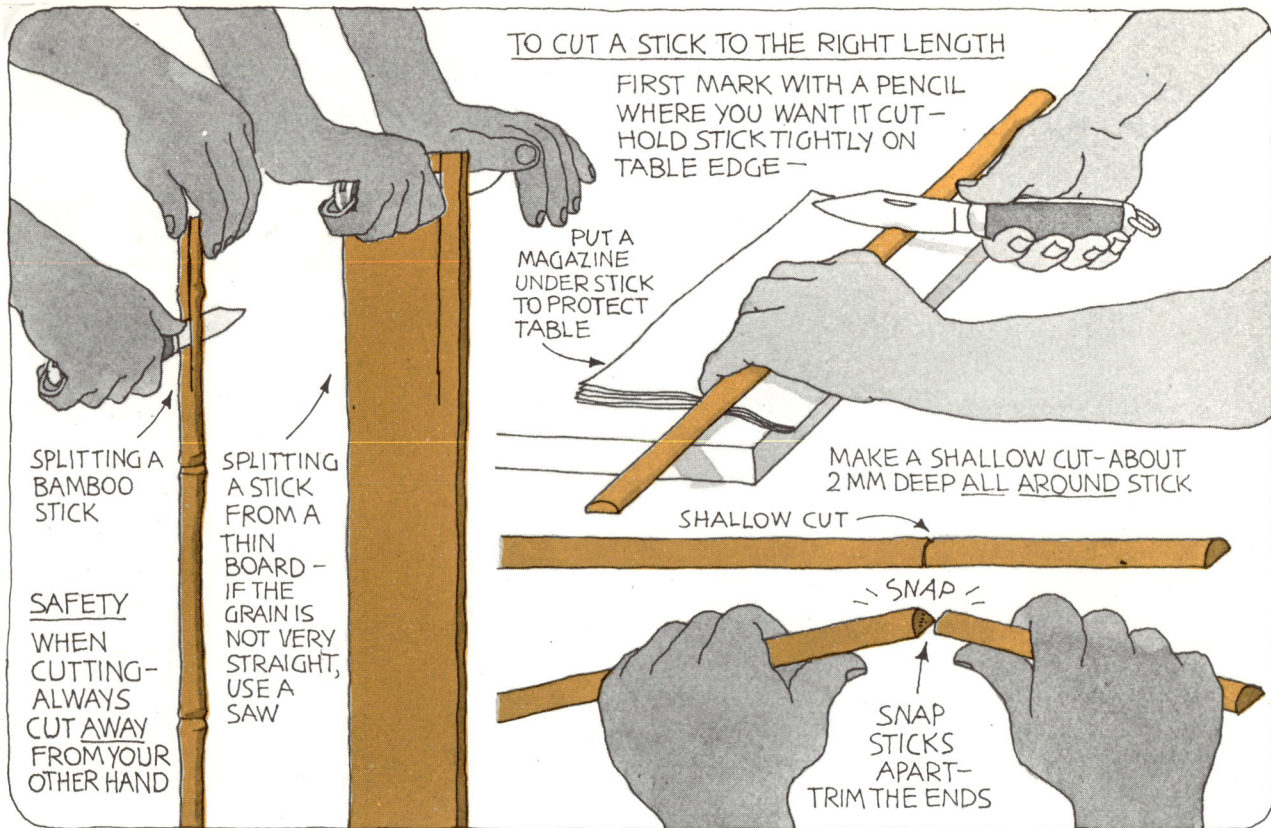

SPLITTING A BAMBOO STICK

SPLITTING A STICK FROM A THIN BOARD — IF THE GRAIN IS NOT VERY STRAIGHT, USE A SAW

SAFETY

WHEN CUTTING — ALWAYS CUT AWAY FROM YOUR OTHER HAND

TO CUT A STICK TO THE RIGHT LENGTH

FIRST MARK WITH A PENCIL WHERE YOU WANT IT CUT — HOLD STICK TIGHTLY ON TABLE EDGE —

PUT A MAGAZINE UNDER STICK TO PROTECT TABLE

MAKE A SHALLOW CUT — ABOUT 2 MM DEEP ALL AROUND STICK

SHALLOW CUT

SNAP

SNAP STICKS APART — TRIM THE ENDS

1 Start by cutting 2 sticks, one 72 cm long, the other 54 cm long.

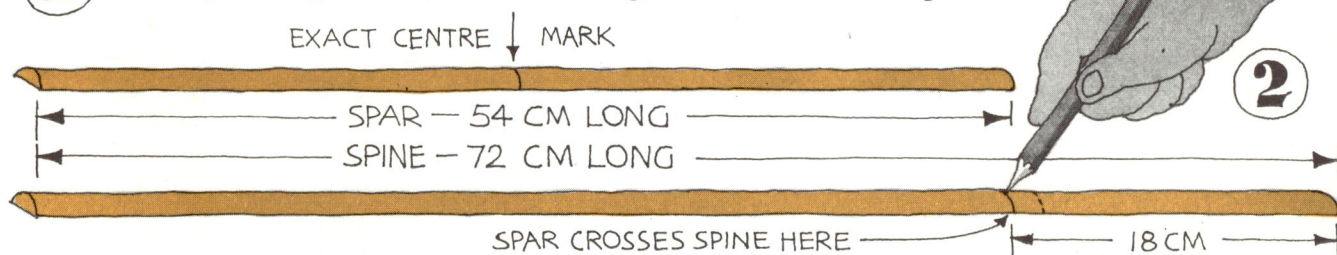

EXACT CENTRE ↓ MARK

SPAR — 54 CM LONG

SPINE — 72 CM LONG

2

SPAR CROSSES SPINE HERE — 18 CM

Find the centre of the short stick (spar) by measuring 27 cm from one end. Mark with a pencil all the way around the stick.

2. Measure 18 cm from one end of the long stick (spine), and mark all the way around. This is where the spar crosses the spine.

(3) NOTCH THE 4 ENDS OF YOUR STICKS — ABOUT 2 MM DEEP — 15MM FROM ENDS

15 MM

IF YOU USE BAMBOO STICKS
NOTCH THEM LIKE THIS

3. Measure 15 mm in from each stick end and cut notches with your knife as shown.

I'D SAY IT WAS A BIT HEAVY THIS END

KNIFE BLADE
CENTRE MARK

IT IS BALANCED ENOUGH IF IT DOESN'T FALL OFF

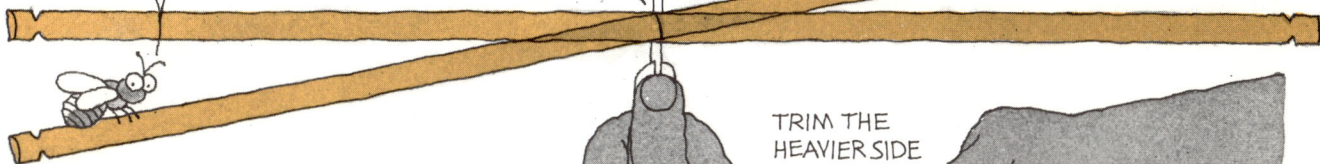

TRIM THE
HEAVIER SIDE

HAND-MADE STICKS
WILL NEED BALANCING

(4) **Balancing the spar**

4. The spar should balance at the centre. Place the stick on the blade of your knife at the centre mark. If it doesn't balance, trim off a little wood towards the heavier end with your knife until it does.

A **bamboo** stick will be thicker at one end than the other, so the heavier side will need a lot of thinning.

TRIM BAMBOO STICKS
ON THE SPLIT SIDE

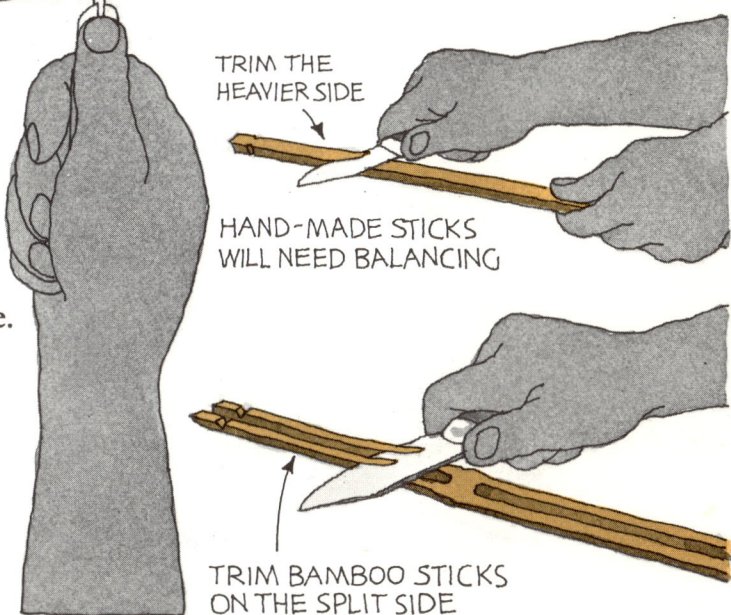

Joining the sticks

Bind the sticks together flat side to flat side. Put a little glue between the sticks before binding.

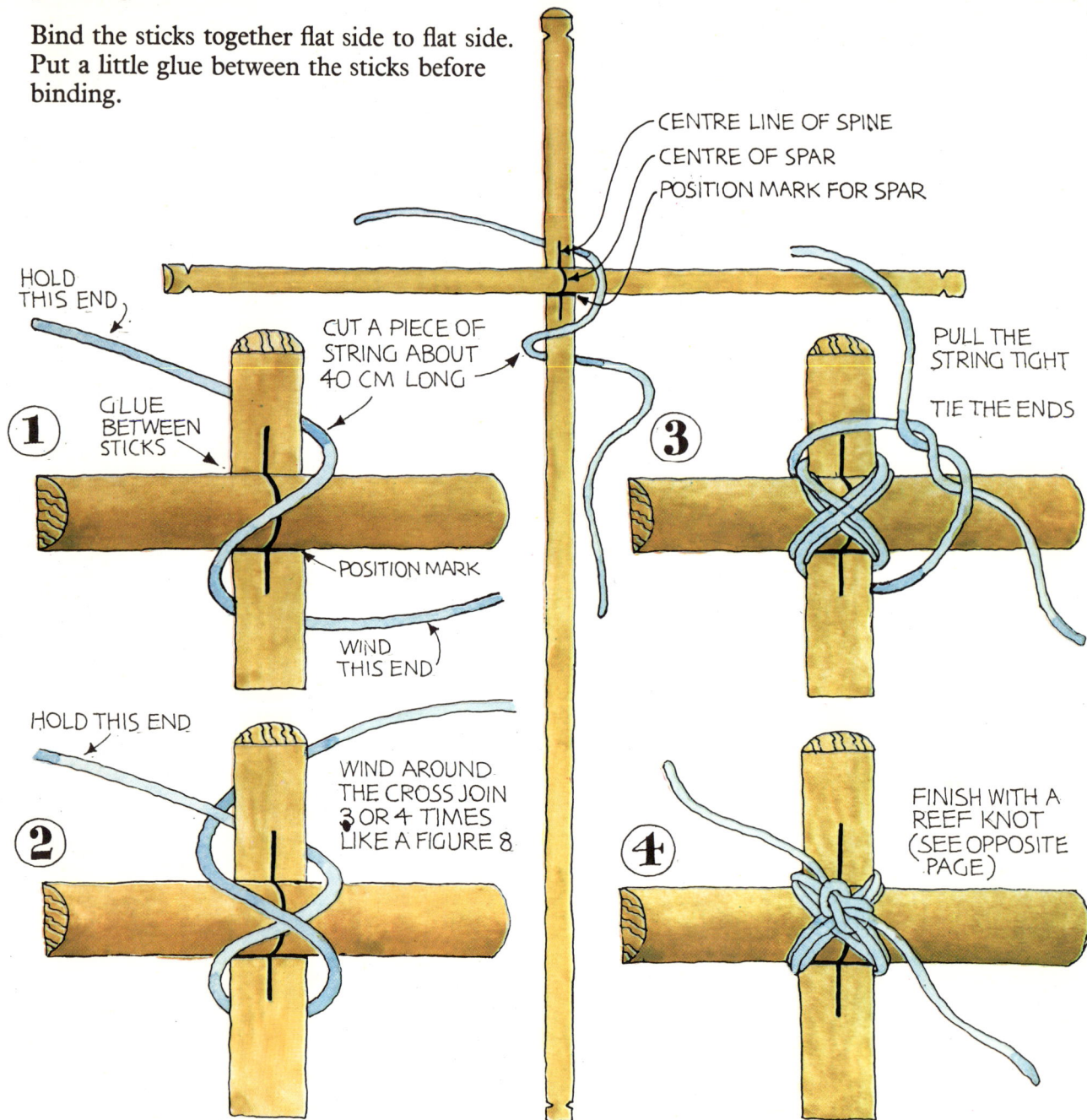

CENTRE LINE OF SPINE
CENTRE OF SPAR
POSITION MARK FOR SPAR

HOLD THIS END

CUT A PIECE OF STRING ABOUT 40 CM LONG

(1) GLUE BETWEEN STICKS

POSITION MARK

WIND THIS END

(3) PULL THE STRING TIGHT

TIE THE ENDS

HOLD THIS END

(2) WIND AROUND THE CROSS JOIN 3 OR 4 TIMES LIKE A FIGURE 8

(4) FINISH WITH A REEF KNOT (SEE OPPOSITE PAGE)

Attaching the tie-strings

First cut 4 pieces of string, each about 30 cm long. These are the tie-strings.

Lie the frame down with the spar on **top**. Tie a tie-string to each notched stick end as shown, with a reef knot facing **up**.

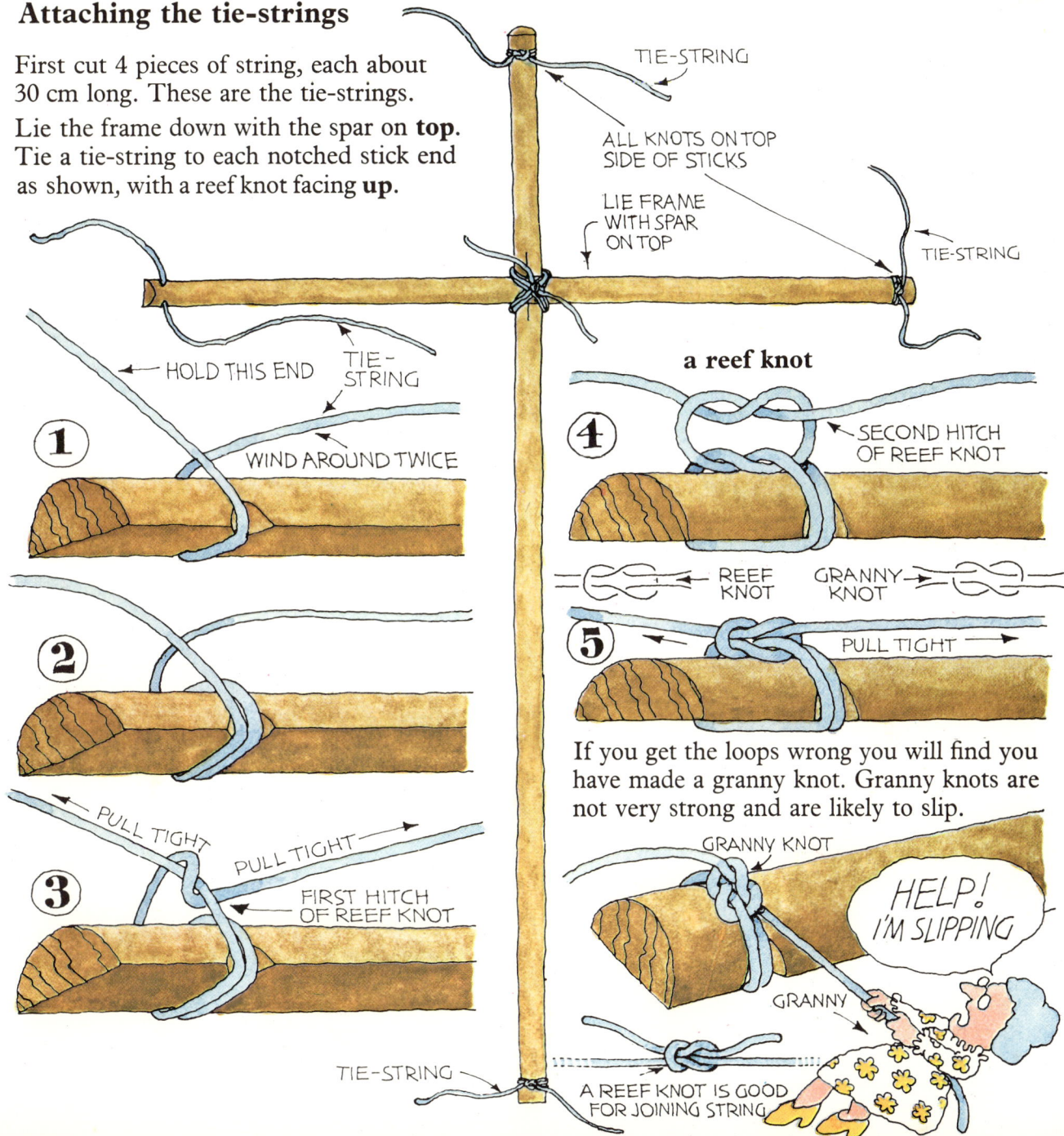

TIE-STRING

ALL KNOTS ON TOP SIDE OF STICKS

LIE FRAME WITH SPAR ON TOP

TIE-STRING

HOLD THIS END

TIE-STRING

① WIND AROUND TWICE

a reef knot

④ SECOND HITCH OF REEF KNOT

REEF KNOT   GRANNY KNOT

⑤ PULL TIGHT

②

③ PULL TIGHT   PULL TIGHT

FIRST HITCH OF REEF KNOT

If you get the loops wrong you will find you have made a granny knot. Granny knots are not very strong and are likely to slip.

GRANNY KNOT

HELP! I'M SLIPPING

TIE-STRING

GRANNY

A REEF KNOT IS GOOD FOR JOINING STRING

The frame-string

1. Cut a piece of string about 2 metres long for the frame-string.

2. Tie a loop in one end.

TOP

TIE FRAME-STRING HERE FIRST

(1)

FRAME-STRING 2 METRES LONG

4 PASS END OF FRAME-STRING TWICE THROUGH LOOP AND PULL TIGHT

PULL

LEFT

RIGHT

HOW TO TIE A LOOP

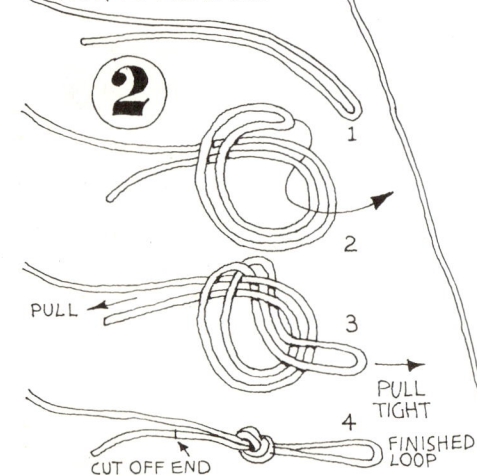

(2)

1

2

PULL

3

PULL TIGHT

4

FINISHED LOOP

CUT OFF END

3

TIE THE FRAME-STRING TO THE STICK ENDS WITH THE TIE-STRING ENDS

FRAME STRING

1. PULL KNOT TIGHT

TIE-STRING ENDS

2. TIE DOWN TIGHTLY WITH REEF KNOT

FRAME STRING

DO NOT CUT OFF THE ENDS YET

3. Tie it to the stick ends with the 4 tie-strings as shown. Tie it to the **top** first, then to the **left** spar end, then the **bottom**, and then the **right** spar end.

4. Complete the frame by passing the free end (twice) through the loop at the other end. You are now ready to tighten and tie the frame-string.

BOTTOM

Tightening the frame-string

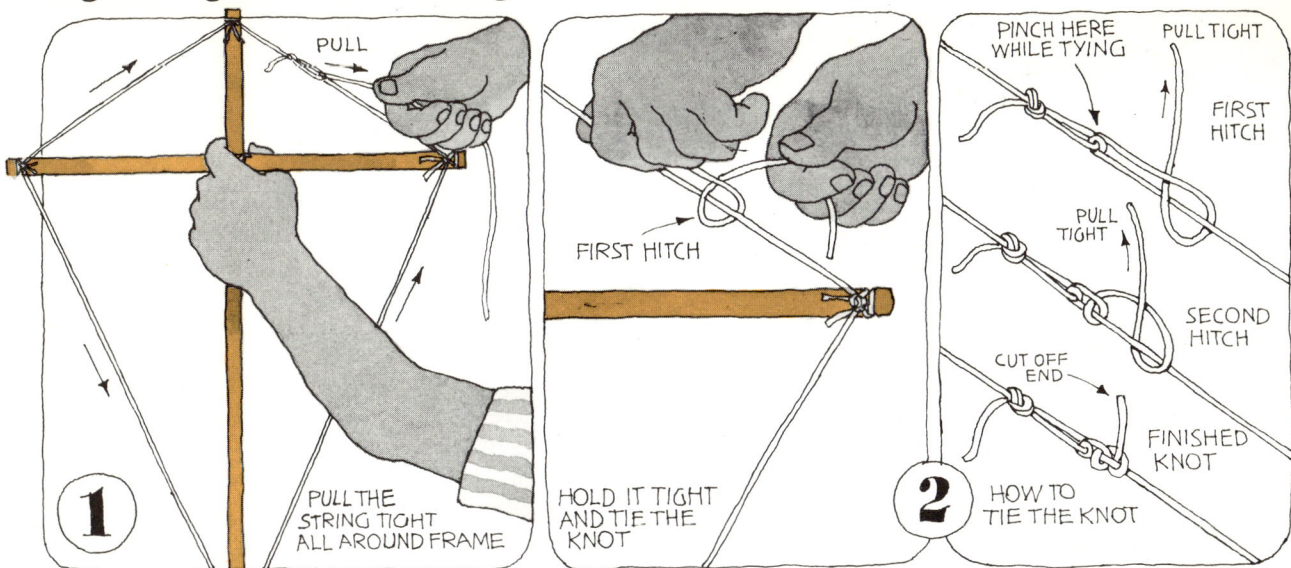

PULL

PULL THE STRING TIGHT ALL AROUND FRAME

(1)

FIRST HITCH

HOLD IT TIGHT AND TIE THE KNOT

PINCH HERE WHILE TYING **PULL TIGHT**

FIRST HITCH

PULL TIGHT

SECOND HITCH

CUT OFF END

FINISHED KNOT

(2) **HOW TO TIE THE KNOT**

1. Grasp the crossed sticks with one hand and pull the frame string tight with the other. Do not pull it so tight that the sticks begin to bend.

2. Hold the string tight by pinching it with finger and thumb where it passes through the loop. Tie it securely with a double hitch as shown.

3. Once the frame is tied, we can check it for squareness. Place a large book in one of the corners at the cross joint. If the square cover of the book doesn't fit snuggly, pull one side of the spar up or down as shown, until it fits. (The tie-strings will allow the sticks to move within the frame.)

4. Stop any further movement by winding the tie-string ends around frame-string and tying. Do this at each corner.

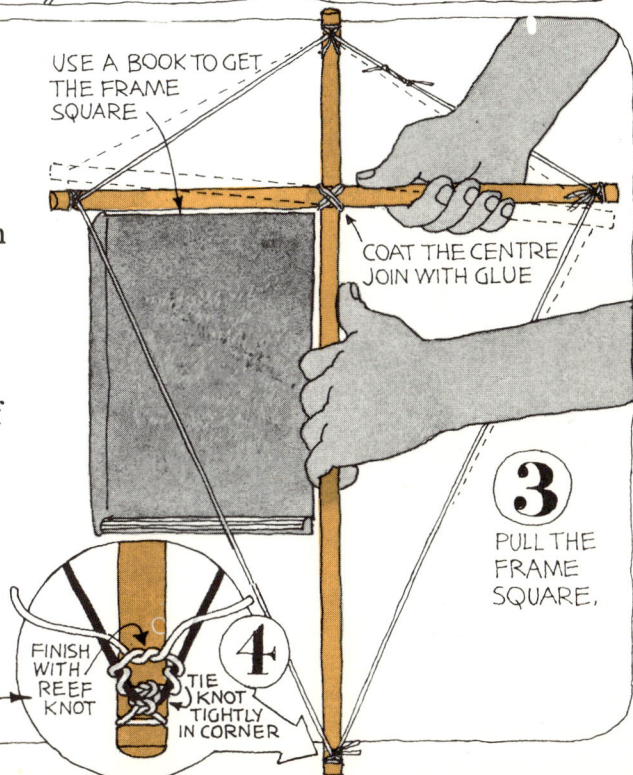

USE A BOOK TO GET THE FRAME SQUARE

COAT THE CENTRE JOIN WITH GLUE

(3) **PULL THE FRAME SQUARE.**

FINISH WITH REEF KNOT

TIE KNOT TIGHTLY IN CORNER

(4)

Covering the frame

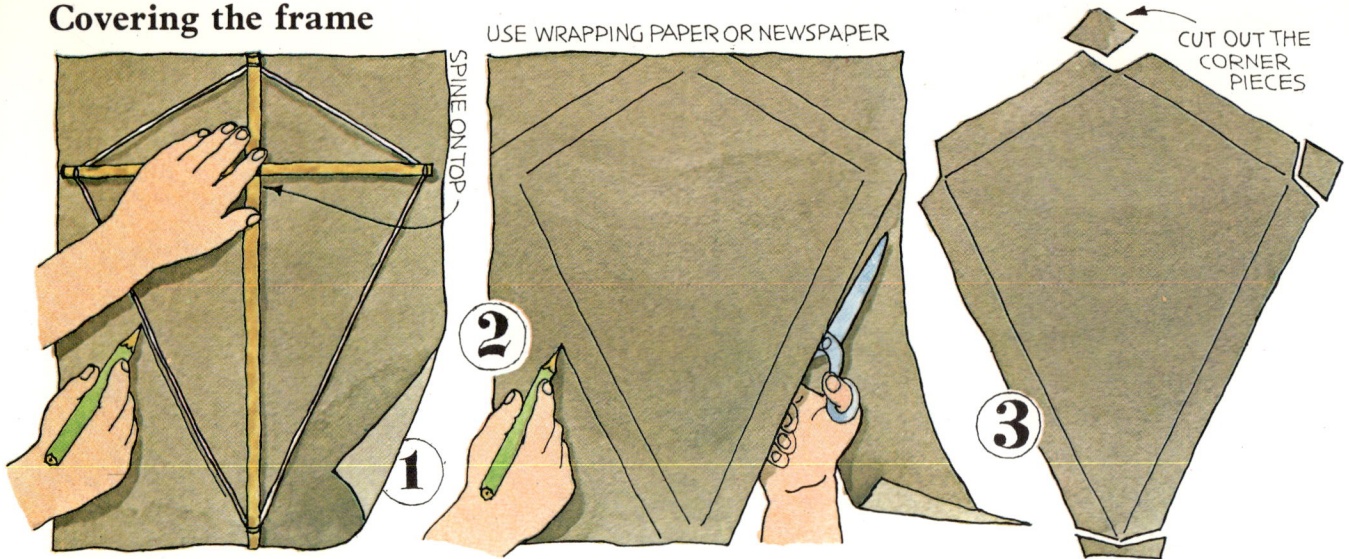

USE WRAPPING PAPER OR NEWSPAPER

SPINE ON TOP

CUT OUT THE CORNER PIECES

1. Place the frame (spine on top) on to a sheet of cover paper. Draw a line around the frame string. **2.** Remove the frame.

Draw another line about 4 cm outside the first line. Cut out the shape along the outside line. **3.** Cut out the corner pieces.

LIE FRAME WITH SPINE ON TOP

STICKY TAPE STRENGTHENS CORNERS

PULL BACK THIS FLAP AND PASTE DOWN OVER STRING

PASTE

TUCK FLAPS UNDER STICKS

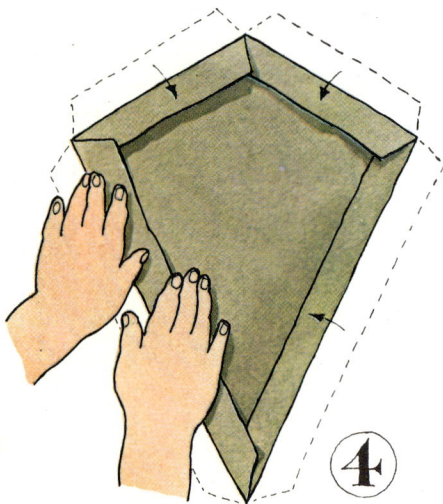

4. Fold the edges in and crease along the inner line. **5.** Replace the frame (spine on top). Place string under folded flaps. Paste down one of the long flaps over the string. Use plenty of

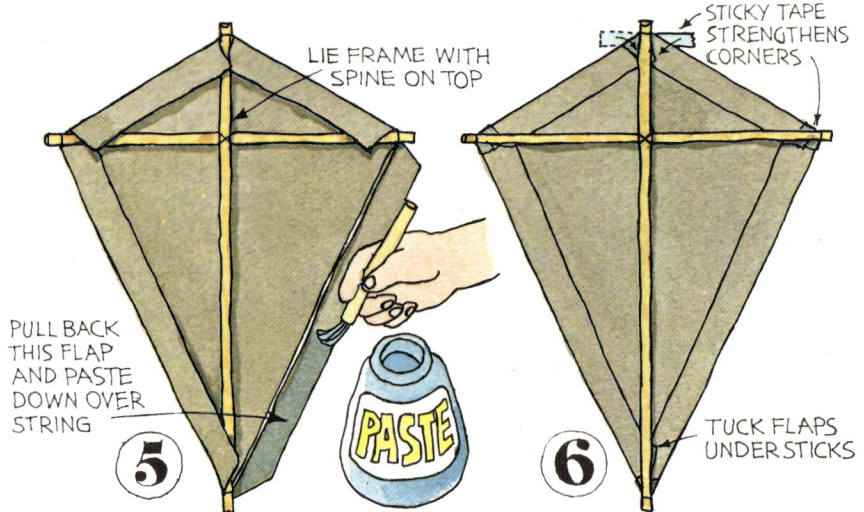

paste, so that the paper becomes limp. **6.** Paste down flap opposite, then the remaining ones. Tuck the ends under the sticks and smooth out.

The rigging – (bridle and tail)

The bridle is used to hold the kite at the correct angle to the wind.

1. Cut a length of string about 1 metre long. Tie one end to the top of the spine.

2. Pass the free end twice through a loop tied at the end of your flying line.

3. Tie the free end of the bridle string to the tail end of the spine allowing about 20 cm slack.

4. Make a mark on the cover about 2 cm below the spar. Make a mark (with a pen) on the bridle string directly opposite this mark. Slide the flying line along the bridle string to this mark. Fix it in place with a piece of sticky tape folded as shown.

A tail keeps the Tadpole the right way up. A paper tail will need to be 4 to 7 times the length of the spine depending on the wind strength. **A.** You can make a tail quickly by tying bunched paper to a tail string.

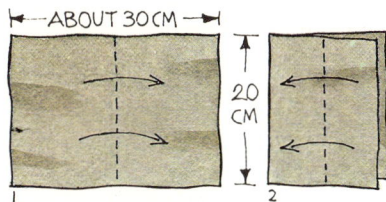

Tie the paper 'feathers' about 30 cm apart until you have a tail about 6 metres long. Tie one end to the tail end of the spine. Cut it off at about 4 metres long. Put the end piece aside for spare. This tail works well but tangles.

B. A 'banner' tail takes longer to make but will not tangle. Cut 35 triangular flags from wrapping paper. Paste them to a tail string as shown. Knotting old stockings or strips of rag together also makes a good tail.

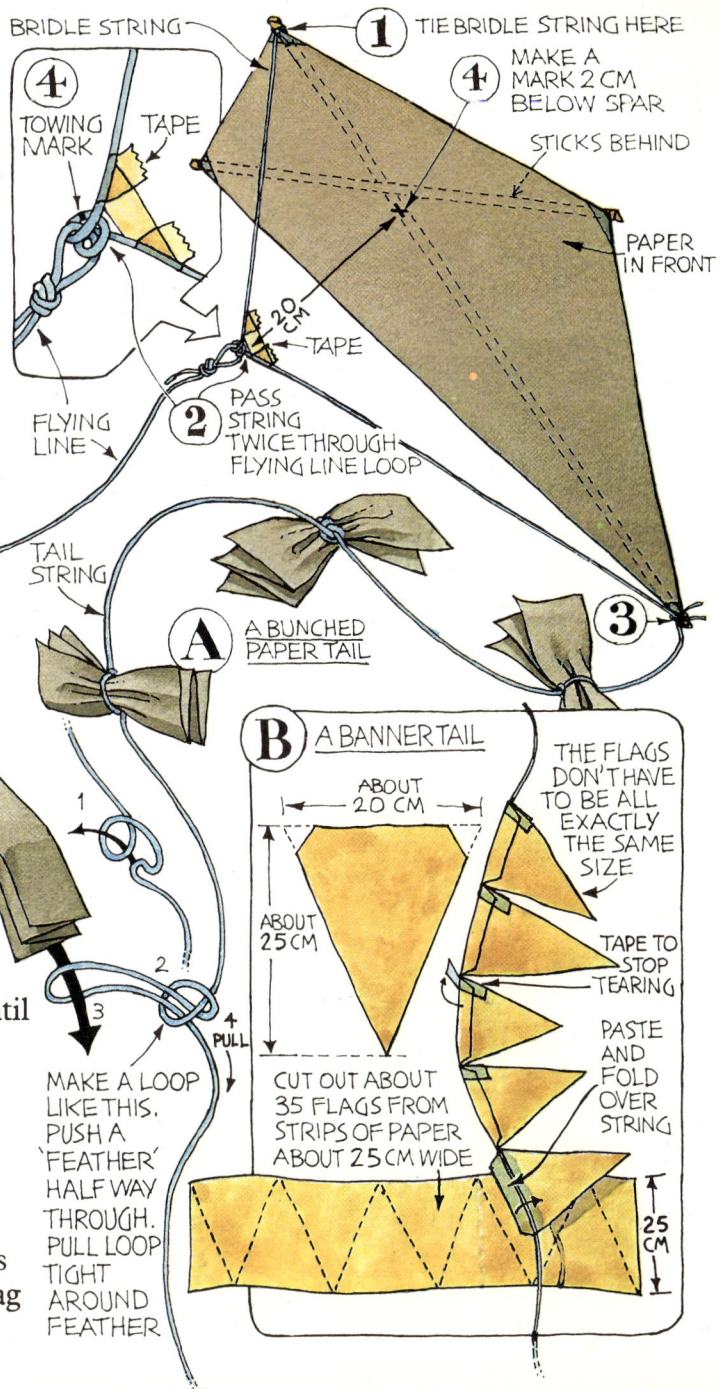

BRIDLE STRING

① TIE BRIDLE STRING HERE

④ MAKE A MARK 2 CM BELOW SPAR

STICKS BEHIND

④ TOWING MARK   TAPE

20 CM   TAPE

PAPER IN FRONT

FLYING LINE

② PASS STRING TWICE THROUGH FLYING LINE LOOP

TAIL STRING

Ⓐ A BUNCHED PAPER TAIL

③

ABOUT 30 CM

20 CM

1   2   3

1

2

3 PULL

4 PULL

MAKE A LOOP LIKE THIS. PUSH A 'FEATHER' HALF WAY THROUGH. PULL LOOP TIGHT AROUND FEATHER

Ⓑ A BANNER TAIL

ABOUT 20 CM

ABOUT 25 CM

CUT OUT ABOUT 35 FLAGS FROM STRIPS OF PAPER ABOUT 25 CM WIDE

THE FLAGS DON'T HAVE TO BE ALL EXACTLY THE SAME SIZE

TAPE TO STOP TEARING

PASTE AND FOLD OVER STRING

25 CM

Getting it in the air

Repair kit: When you go out flying take a first-aid kit for quick repairs (to the kite, not to you). Take your pocket knife, scissors, a roll of sticky tape, paper for patching and 2 metres of extra string. Most important, with the tadpole, is to have plenty of extra tail material in case you have to extend it. In an emergency you can tie on your handkerchief (or your socks).

Take this book to remind you how to adjust your kite. Put these things in a carrying bag. Remember to put them back in the bag after using them—it is easy to lose a small pocket knife in a large field.

Launching: Stand in the field, so that the kite will be clear of power lines, trees, and people. Open parks and beaches are good places to fly.

Launching with a friend

1. Your friend holds the kite up facing the wind like this, ready to release.

2. Walk about 40 paces from kite, **into** the wind. Unwind your line as you go.

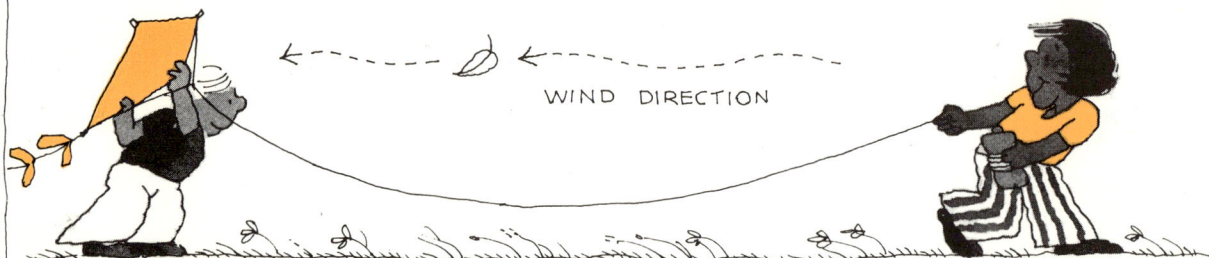

WIND DIRECTION

3. Turn and face the kite. Hold the reel tightly. Take up the slack by walking backwards, pulling the kite from your friend's hands. Feed it more line as it climbs. Give 3 hearty cheers.

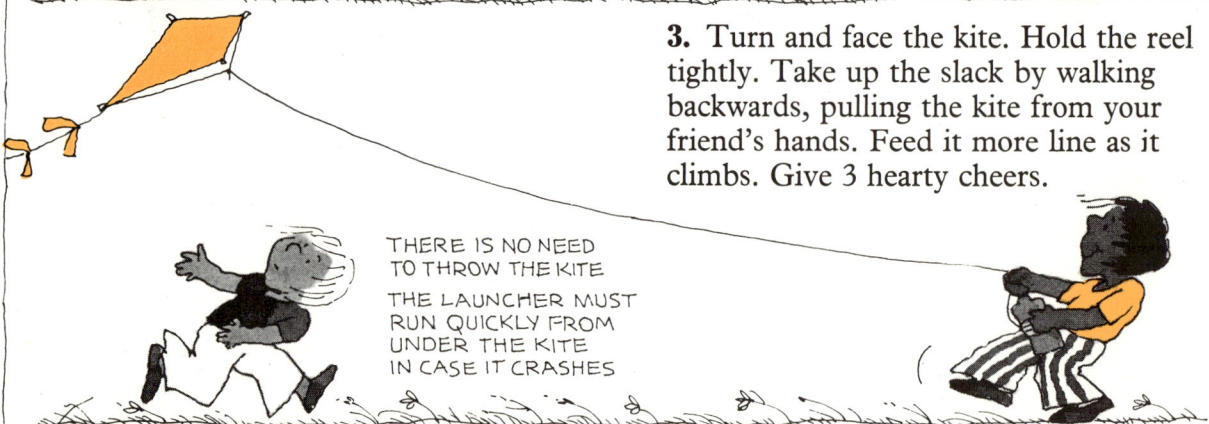

THERE IS NO NEED TO THROW THE KITE

THE LAUNCHER MUST RUN QUICKLY FROM UNDER THE KITE IN CASE IT CRASHES

If the wind is not very strong, run into the wind to get it up. There may be enough wind higher up but if not, no amount of galloping around will keep it up. Wait for a windier day.

Launching it by yourself

1. Stand with your back to the wind. Hold the kite up to catch the wind.

WIND DIRECTION

2. As the wind takes it up let out more line. Don't let it run too quickly through your fingers –it cuts.

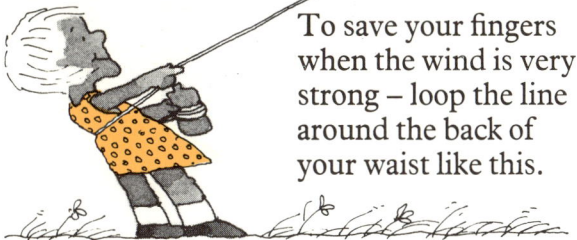

To save your fingers when the wind is very strong – loop the line around the back of your waist like this.

To stop a wild kite from crashing

Run forward with the reel, or let it fall to the ground. Your kite will come down gently.

Another way to launch it

You can get your kite up with the line out, by pulling it along the ground into the wind. Lie it face down, nose forward. Prop the nose up a bit with a stick or tuft of grass so that the wind can get under it. Don't let the kite get wet.

WIND DIRECTION
PROP UP NOSE A LITTLE
LIE TAIL AHEAD OF KITE

Adjusting the rigging

There are 2 things that kites do wrong. 1. Dive and spin. 2. Refuse to climb.

They can be adjusted to make them fly better. Tailed kites can be adjusted at 2 places – the tail and the bridle.

1. The tail: The stronger the wind the longer the tail needed to keep it steady.

2. The bridle: By moving the towing position up or down the bridle string you can change the kite's 'angle of attack' to the wind. The stronger the wind the higher the towing position.

Diving and spinning can be corrected by making the tail longer. But this adds weight and 'drag'. It will fly better and higher if we can correct it by adjusting the bridle.

Remove the sticky tape and slip the flying line loop **down** the bridle string about 2 cm towards the tail. The mark on the bridle string will help you to judge where to make the new position. Fix it in place with a fresh piece of tape and try again. If you move it down too far the kite will not rise.

If the tail is long enough you will find a position where the kite will fly well. But if it is still unsteady, tie on an extra metre or two of tail and move the line back to the first position.

If your kite **will not climb** very high its tail could be too long for the wind that is blowing. Or the flying line may need to be moved **up** the bridle string a bit. Experiment with tail and bridle until your kite is flying high.

The Bat – a bowed kite

The Bat does not have a tail to wag (or to get all tangled up). It is able to fly because it presents a **curved** surface to the wind – not flat like that of the Tadpole.

It has a crossed stick frame – made in the same way as the Tadpole's, but its spar is much longer than its spine. The spar is made to curve by tying the ends back with a bowstring – like an archer's bow.

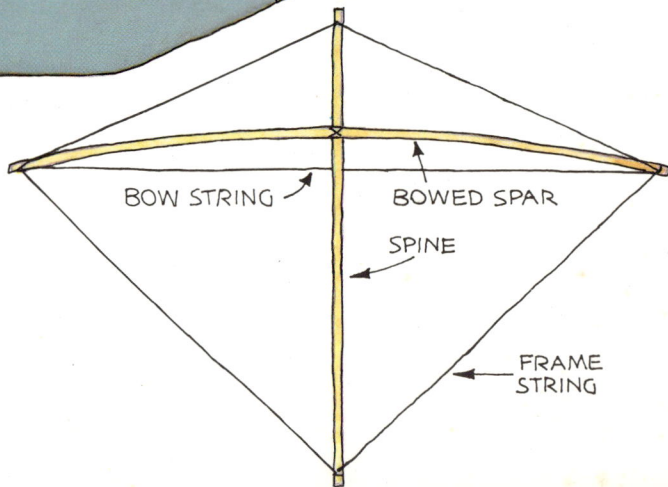

Bowing the spar

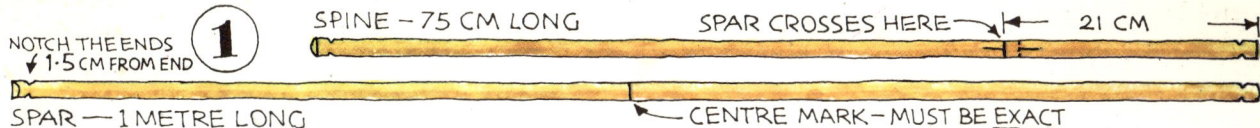

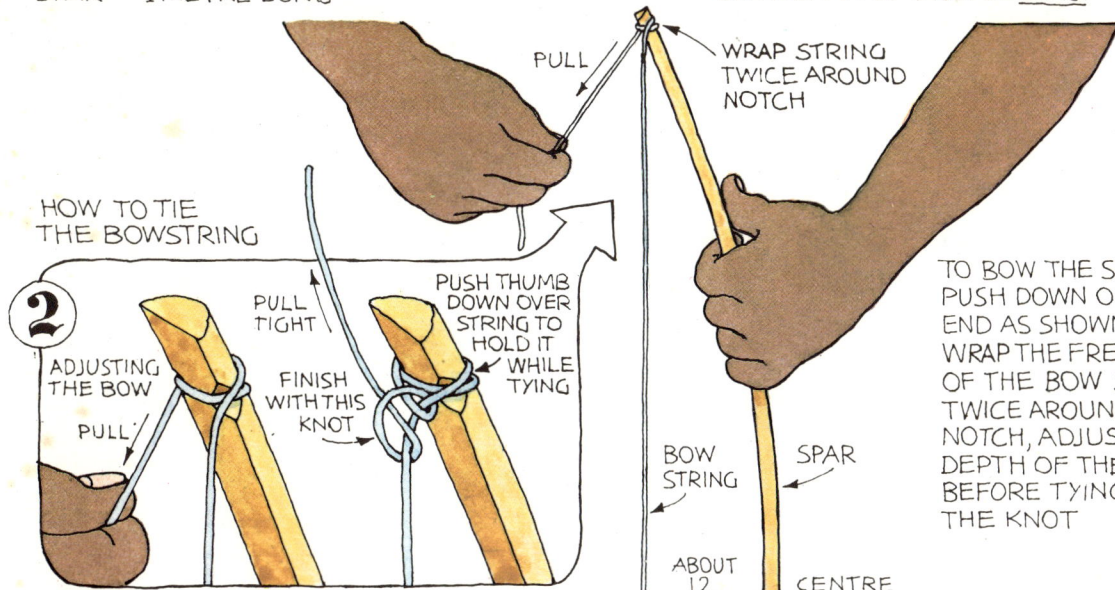

NOTCH THE ENDS
↓ 1·5 CM FROM END
(1) SPINE – 75 CM LONG

SPAR CROSSES HERE 21 CM

SPAR – 1 METRE LONG

CENTRE MARK – MUST BE <u>EXACT</u>

PULL

WRAP STRING
TWICE AROUND
NOTCH

HOW TO TIE
THE BOWSTRING

(2) PULL
TIGHT

PUSH THUMB
DOWN OVER
STRING TO
HOLD IT
WHILE
TYING

ADJUSTING
THE BOW

FINISH
WITH THIS
KNOT

PULL

BOW
STRING

SPAR

ABOUT
12
CM

CENTRE

FLAT
SIDE

TO BOW THE SPAR
PUSH DOWN ON THIS
END AS SHOWN.
WRAP THE FREE END
OF THE BOW STRING
TWICE AROUND THE
NOTCH, ADJUST THE
DEPTH OF THE BOW
BEFORE TYING
THE KNOT

1. The sticks: Cut 2 sticks from fly bead moulding or use other sticks. Make the spar 1 metre long and the spine 75 cm long. Check that the spar will bend a bit without breaking. Cut notches in the ends. Mark the centre and balance the spar.

2. Cut a bow-string about 130 cm long. Tie one end to the spar as shown, with the knot on the flat side.

Bend the spar and loop the bow-string twice around the other end of the spar. Adjust the depth of the bow to about 12 cm before tying the knot.

TIE THE STRING TO THIS END FIRST
PLACE THIS END ON THE GROUND
AND PUSH DOWN ON THE OTHER END

Note: It is important that the spar bends fairly evenly on both sides of the centre. Bought moulding sticks usually bend evenly. A split bamboo stick will need a lot of trimming on one side to get it to bow correctly.

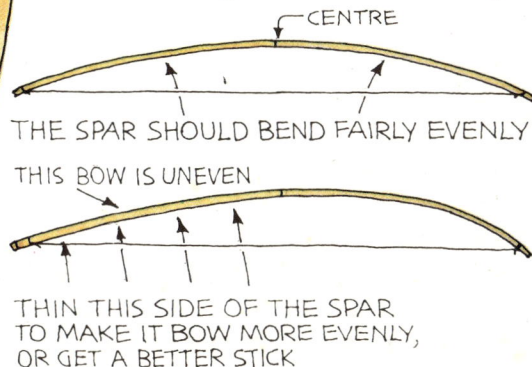

CENTRE

THE SPAR SHOULD BEND FAIRLY EVENLY

THIS BOW IS UNEVEN

THIN THIS SIDE OF THE SPAR
TO MAKE IT BOW MORE EVENLY,
OR GET A BETTER STICK

Finishing the frame

Note: Have your spar bowed before starting the frame.

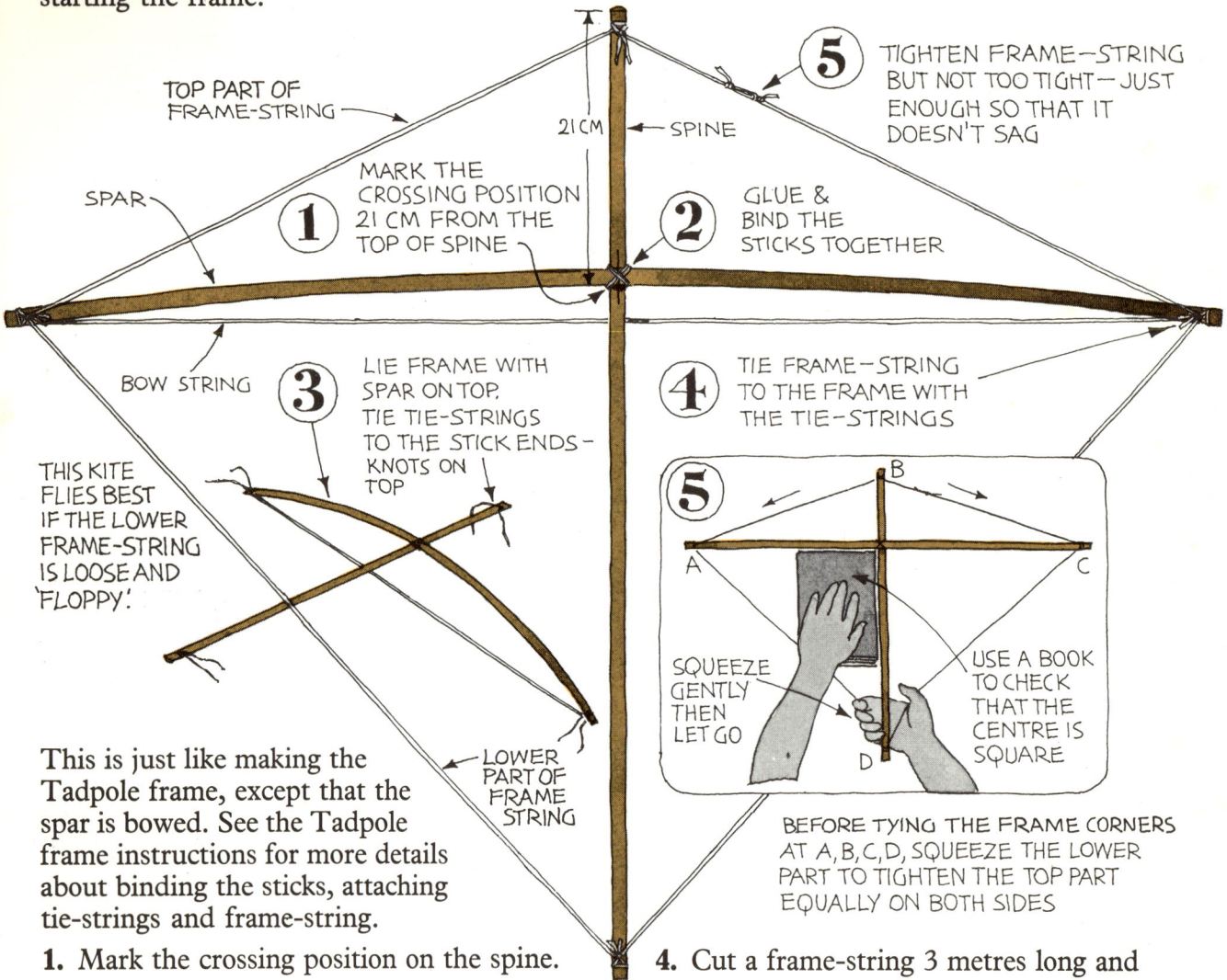

TOP PART OF FRAME-STRING

SPAR

BOW STRING

21 CM ← SPINE

1 MARK THE CROSSING POSITION 21 CM FROM THE TOP OF SPINE

2 GLUE & BIND THE STICKS TOGETHER

5 TIGHTEN FRAME-STRING BUT NOT TOO TIGHT—JUST ENOUGH SO THAT IT DOESN'T SAG

3 LIE FRAME WITH SPAR ON TOP, TIE TIE-STRINGS TO THE STICK ENDS— KNOTS ON TOP

4 TIE FRAME-STRING TO THE FRAME WITH THE TIE-STRINGS

THIS KITE FLIES BEST IF THE LOWER FRAME-STRING IS LOOSE AND 'FLOPPY'.

LOWER PART OF FRAME STRING

5 SQUEEZE GENTLY THEN LET GO

USE A BOOK TO CHECK THAT THE CENTRE IS SQUARE

BEFORE TYING THE FRAME CORNERS AT A, B, C, D, SQUEEZE THE LOWER PART TO TIGHTEN THE TOP PART EQUALLY ON BOTH SIDES

This is just like making the Tadpole frame, except that the spar is bowed. See the Tadpole frame instructions for more details about binding the sticks, attaching tie-strings and frame-string.

1. Mark the crossing position on the spine.

2. Glue and bind the sticks together (flat side to flat side).

3. Lie the frame down, **spar** facing up. Tie tie-strings to the stick ends (knots facing up).

4. Cut a frame-string 3 metres long and tie a loop in one end. Tie it to the stick ends with the tie-strings.

5. Tighten the frame-string. Square the cross joint and tie the frame-string corners.

You are now ready for the cover.

Covering the Bat frame

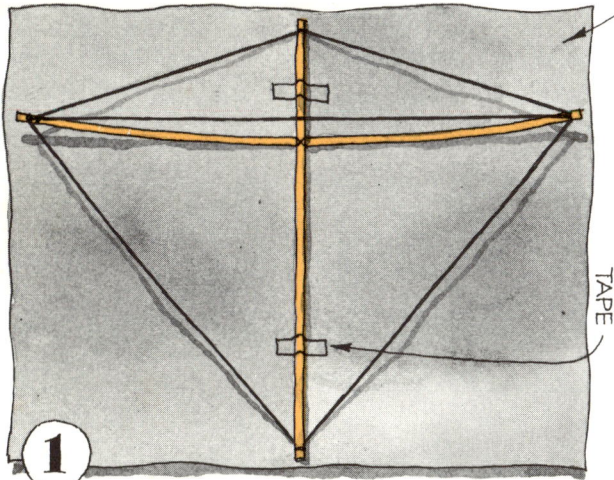

WRAPPING PAPER

TAPE

TAPE

ABOUT 6 CM

1. Lie the frame (bow-string facing up) on a piece of cover paper (if you use newspaper, join 2 double sheets). Allow plenty of overlap for shaping. Tape the spine to the paper.

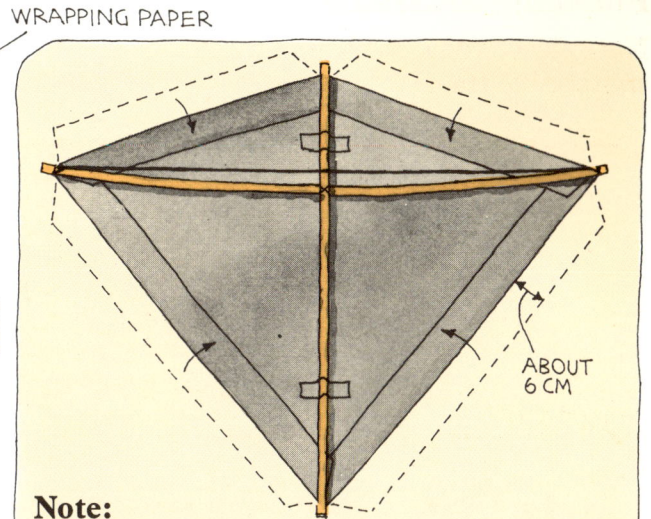

Note:
If you don't wish to make your kite with a Bat shape, cut seams - paste them over the string like the Tadpole cover. It will fly just as well as a kite with a shaped cover.

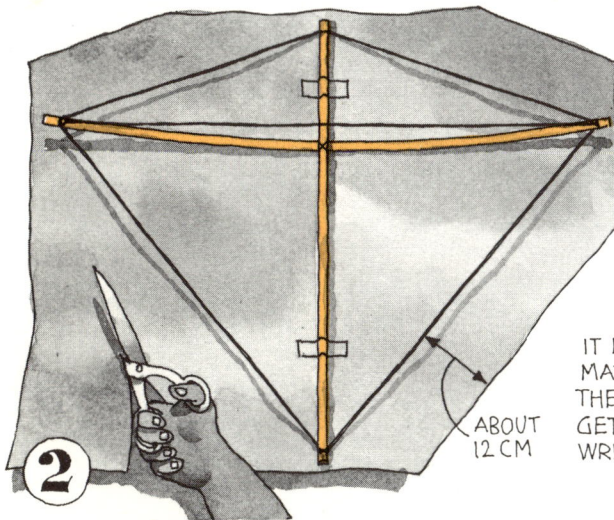

ABOUT 12 CM

IT DOESN'T MATTER IF THE PAPER GETS A BIT WRINKLED

PUSH PAPER (OR TAPE) FIRMLY DOWN OVER STRING

2. To make the Bat shape, first cut off the corners about 12 cm from the string.
3. Paste strips of paper (about 30 cm long 6 cm deep) over the string onto the cover all around the frame. Overlap the strips. First stick down the long, then the short sides on the left. Then the long and short sides on the right. Let the paste dry before doing the next thing.

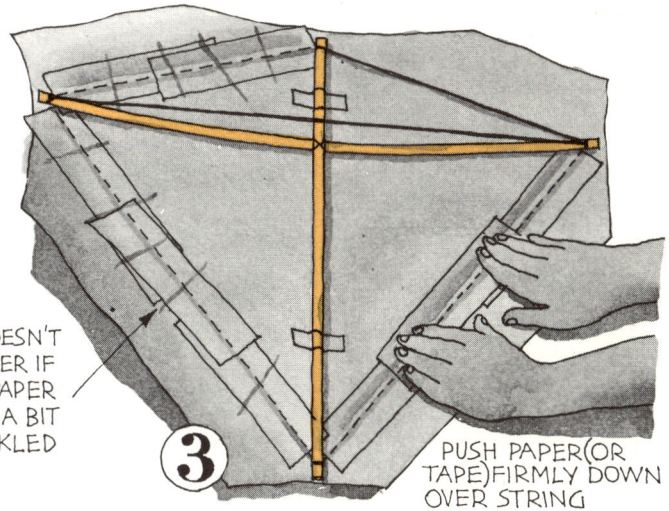

Shaping the cover

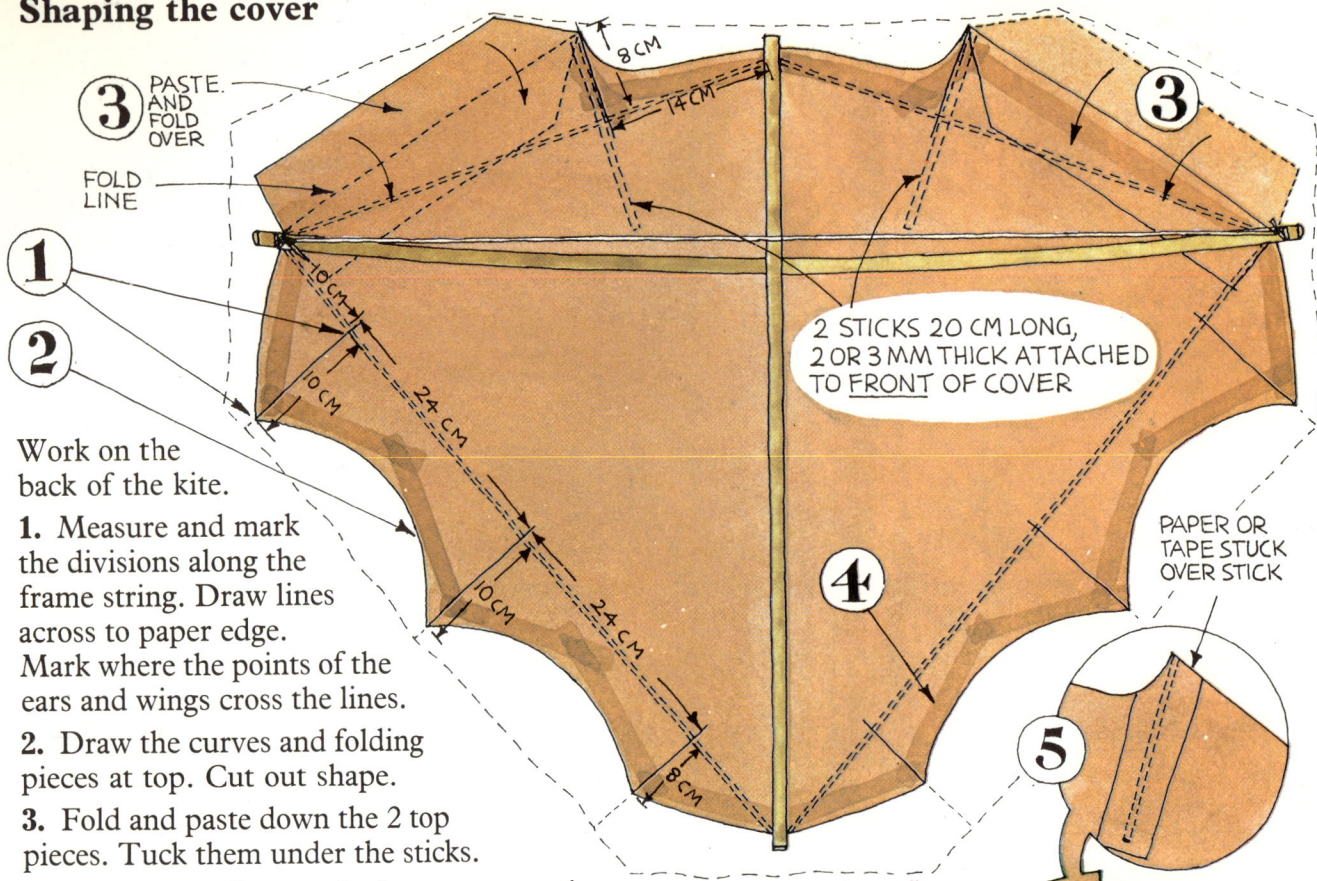

③ PASTE AND FOLD OVER

FOLD LINE

① ②

8 CM

14 CM

10 CM

10 CM

24 CM

10 CM

24 CM

8 CM

③

④

2 STICKS 20 CM LONG, 2 OR 3 MM THICK ATTACHED TO FRONT OF COVER

PAPER OR TAPE STUCK OVER STICK

⑤

Work on the back of the kite.

1. Measure and mark the divisions along the frame string. Draw lines across to paper edge. Mark where the points of the ears and wings cross the lines.

2. Draw the curves and folding pieces at top. Cut out shape.

3. Fold and paste down the 2 top pieces. Tuck them under the sticks.

4. Sticky tape all around edge to stop tearing.

5. Turn the kite over. Tape or paste thin sticks 20 cm long, 2 or 3 mm thick to each ear. This stops them bending back in the wind. Thinly split bamboo or straight twigs from a tree will do.

6. Paint the front of the kite all over with black paint (poster or acrylic). Don't worry if it wrinkles.

7. From a piece of white paper, cut out 2 circles about 15 cm across, for eyes. Cut out 5 triangular pieces for teeth. Make the 2 outside teeth about 12 cm long, the others about 8 cm. Paste teeth and eyes to the kite to make a face, something like the drawing. Paint in eye dots.

⑥

⑦

The bridle

MAKE A MARK 4 CM BELOW SPAR

25 CM

TAPE

BRIDLE STRING

MARK STRING HERE

FLYING LINE

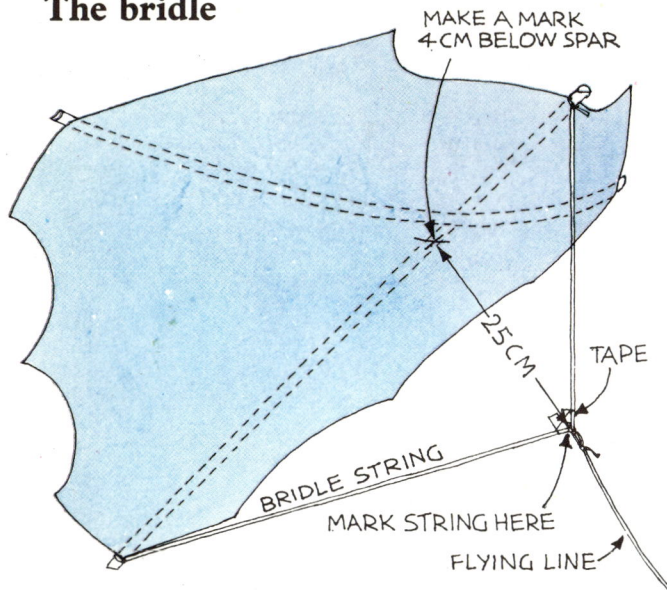

The Bat's bridle is similar to the Tadpole's.

1. Cut a bridle string 1.5 metres long.

2. Tie one end to the top of the spine.

3. Pass the free end of the bridle string twice through the loop at the end of your flying line. Tie the free end to the bottom of the spine, allowing about 25 cm slack.

4. Make a mark on the cover 4 cm below the spar. Hold the bridle string taut at a point opposite the mark and mark the bridle string (with a pen) at this position. Fix the flying line loop to this mark.

Flying the Bat

Launch the Bat like the Tadpole.

Adjustments: You have 2 places to make adjustments, the **bridle** and the **bow**.

Bridle adjustments are made in the same way as on the Tadpole. In very light winds the line can be moved up a bit. This helps the kite to climb higher. But in strong winds move it down again or it will spin.

The bow depth can be increased to about 15 cm for steady flying in very strong winds.

You may find, after trying different bridle positions, that the Bat won't climb and keeps pulling and diving to one side – always the same side. This means that the side of the kite that dips down is catching more wind than the other side. It needs **wind balancing.** We must make it so that the wind pushes **equally** on both sides.

HOLE FOR SPAR

ATTACH STREAMER TO THIS SIDE

ZOOM!

TAPE

STREAMER

DIVING SIDE

Cut a paper 'streamer' about 25 cm long, 5 cm wide. Make a hole for the spar. Tape it to the side that tips up. The wind will flutter the streamer giving extra push to this side, balancing the push on the other side. You may have to use a longer or shorter streamer to get it right. Place sticky tape down the centre of the strip and around spar hole to stop tearing.

The Minuteman 'Stunt' Kite

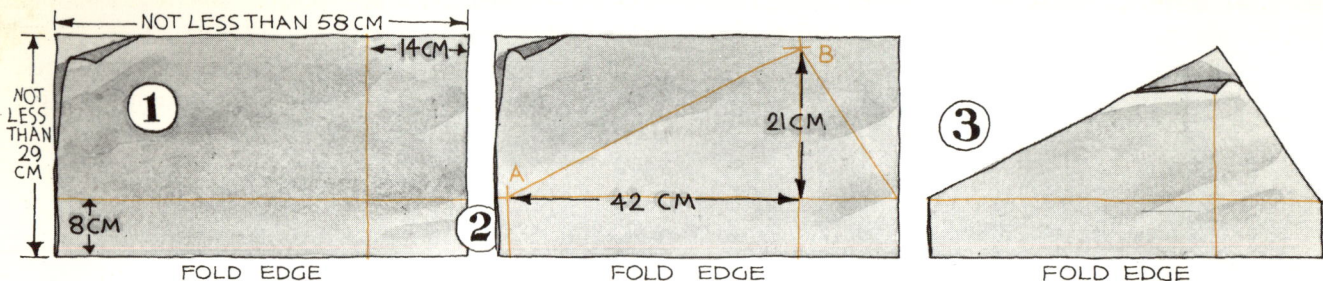

1 NOT LESS THAN 58 CM — 14CM
NOT LESS THAN 29 CM
8CM
FOLD EDGE

2 B, 21CM, A, 42 CM
FOLD EDGE

3
FOLD EDGE

A Minuteman Kite is quick to make – no pasting.
1. Find a sheet of newspaper or wrapping paper. It can be larger than 58 cm square, but not smaller. Bring 2 opposite sides together folding it in half. Crease it. Lie the doubled sheet on the table, folded side towards you. Measure 14 cm in from right hand side. Rule a line level with the side. Measure 8 cm up from the fold. Rule a line (level with fold) crossing first line.
2. Measuring from where the lines cross, mark points A and B. Rule the three lines shown.
3. Cut out the shape, still folded.

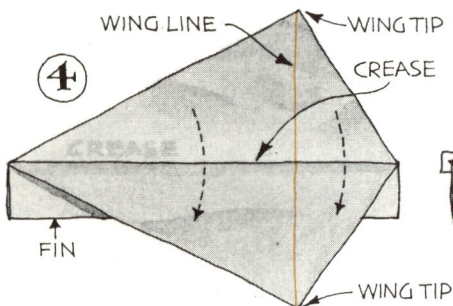

4 WING LINE — WING TIP — CREASE — CREASE — FIN — WING TIP

5

6 TAPE

4. Fold the top sheet down along the line and crease. Rule the 'wing line' from corner to corner.
5. Join the 'wings' with a strip of tape along the crease. The folded piece underneath is the 'fin'.
6. Stick strips of tape along the wing edges, (on one side of paper only) to stop tearing.

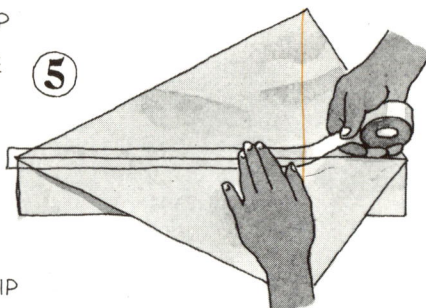

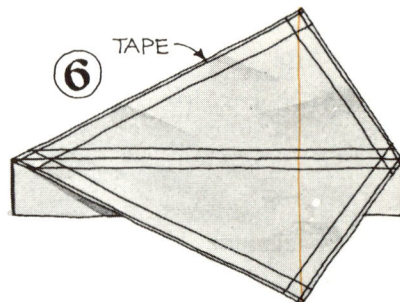

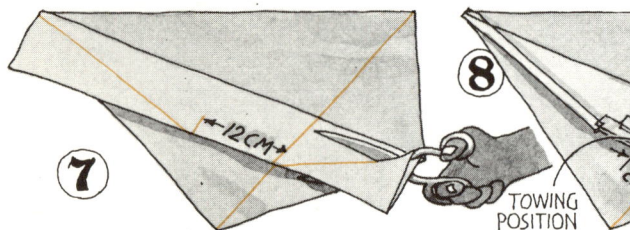

7 ←12CM→

8 3.5 CM — TOWING POSITION

9 FOLD TAPE UNDER — CARDBOARD — 35 CM — 4 CM — MARK — 8CM

7. Turn kite over. Rule wing line across fin. Measure 12 cm back from the wing line. Rule the 2 lines – cut off fin corners here.
8. Mark towing position 3.5 cm back from wing line. Stick 2 strips of tape over mark.
Tape double edges of fin together as shown.
9. Cut a strip of thin cardboard (from a breakfast cereal box) 35 cm long, 4 cm deep. Mark 8 cm from one end. Cut off 2 corners. Tape it to fin 1 cm below wing crease, with 8 cm mark over wing line.

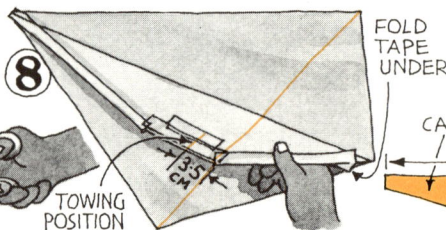

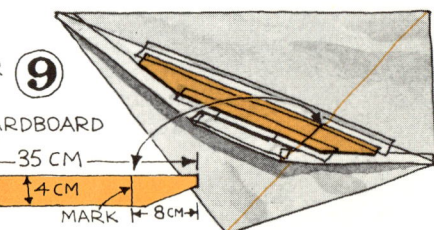

CENTRE MARK

CUT A THIN 'SPRINGY' STICK 44 CM LONG

TRIM IT TO ABOUT 3MM THICK ⟶

ACTUAL SIZE DRAWING

(10) THE MORE EVENLY IT BOWS THE BETTER THE KITE FLIES

IT SHOULD BEND EASILY. MAKE IT THINNER TOWARDS THE ENDS

TAPE THE STICK DOWN AT THE CENTRE FIRST

(11)

STICK TO WING TIP

3 CM

WIND 2 TURNS OF TAPE ONTO BOTH ENDS OF STICK

STICK TO KITE

WING LINE

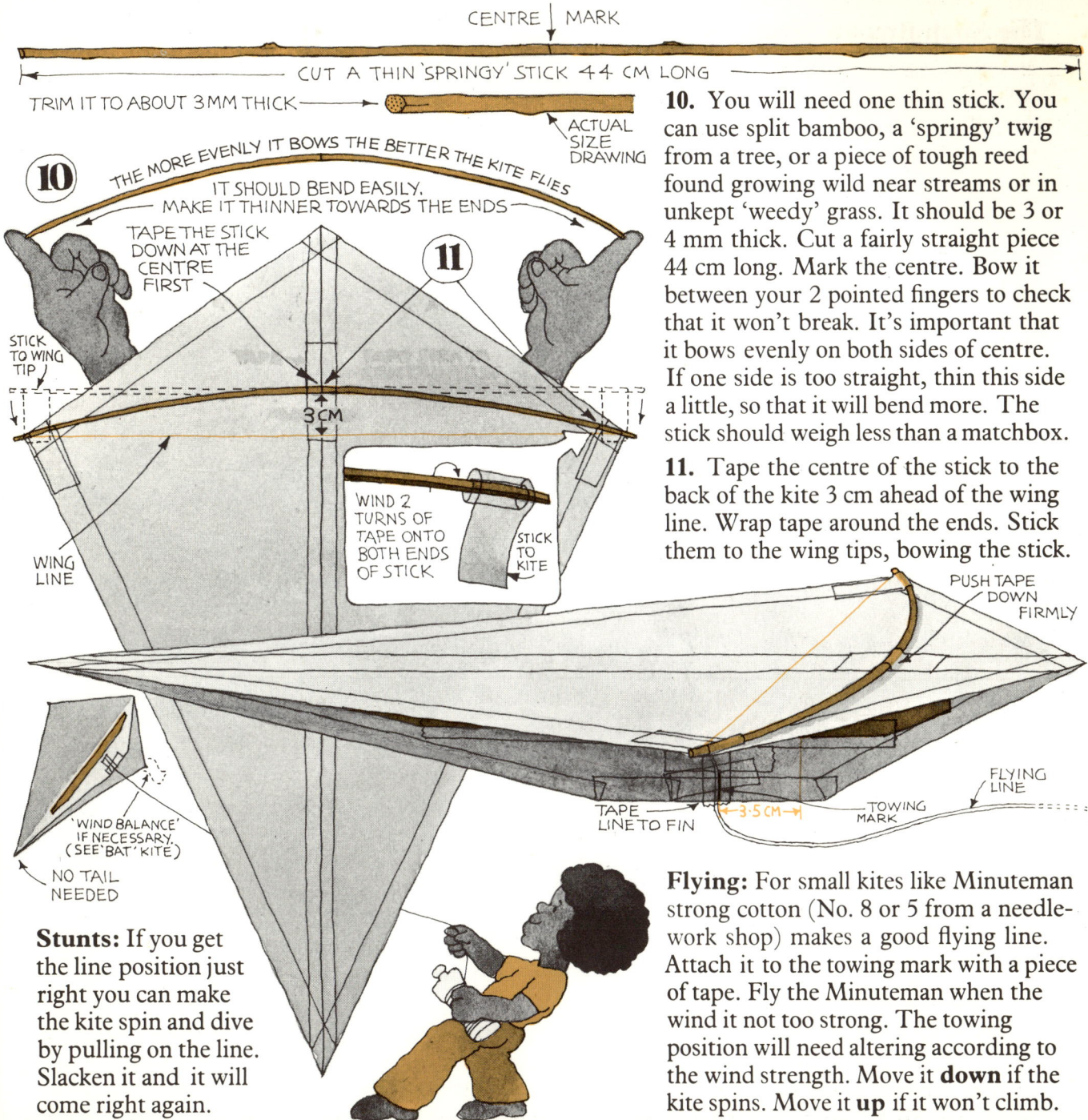

10. You will need one thin stick. You can use split bamboo, a 'springy' twig from a tree, or a piece of tough reed found growing wild near streams or in unkept 'weedy' grass. It should be 3 or 4 mm thick. Cut a fairly straight piece 44 cm long. Mark the centre. Bow it between your 2 pointed fingers to check that it won't break. It's important that it bows evenly on both sides of centre. If one side is too straight, thin this side a little, so that it will bend more. The stick should weigh less than a matchbox.

11. Tape the centre of the stick to the back of the kite 3 cm ahead of the wing line. Wrap tape around the ends. Stick them to the wing tips, bowing the stick.

PUSH TAPE DOWN FIRMLY

FLYING LINE

TAPE LINE TO FIN

3.5 CM

TOWING MARK

'WIND BALANCE' IF NECESSARY. (SEE 'BAT' KITE)

NO TAIL NEEDED

Stunts: If you get the line position just right you can make the kite spin and dive by pulling on the line. Slacken it and it will come right again.

Flying: For small kites like Minuteman strong cotton (No. 8 or 5 from a needlework shop) makes a good flying line. Attach it to the towing mark with a piece of tape. Fly the Minuteman when the wind it not too strong. The towing position will need altering according to the wind strength. Move it **down** if the kite spins. Move it **up** if it won't climb.

The Sock Box – a box kite

WHERE DO I PUT ALL THESE SOCKS?

COVER

CORNER STICK

BRIDLE

STRUTS

COVER

Younger kite makers may need help from an older person to make this one.

The Sock Box is a square box kite. It looks like 2 boxes joined with sticks. But of course they are not really boxes, but square tubes called cells – open at the ends to allow the air to flow through.

This is how it is made. First you make the 2 cell covers from strong wrapping paper. 4 long sticks are placed inside the cells at the corners. The corner sticks are spread with tight-fitting bracing struts. These struts keep the covers taut and make the kite rigid.

Making the cell covers

36 CM

DEPTH CAN BE MORE BUT NOT LESS THAN 50 CM

CREASE

1

NOT LESS THAN 160 CM

CREASE

2

CREASE

3

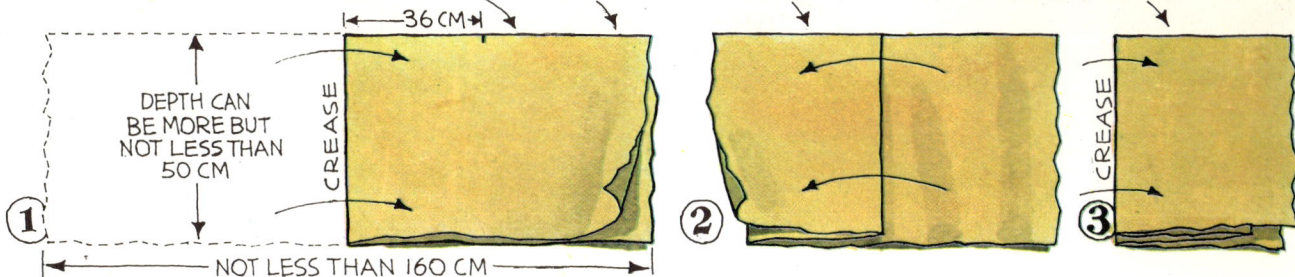

Ask at a shop for a piece of wrapping paper from a roll. It must be at least 50 cm wide and 160 cm long. If you can't get one long piece, paste 2 or 3 small pieces together to make the length. The top edge must be straight.

1. Fold it in the centre. Hold the top edges together when creasing. Measure and mark 36 cm from crease (on top and bottom sheets.)
2. Fold back top sheet at mark and crease.
3. Fold back bottom sheet at mark and crease.

41 CM

4

5

10 CM

6

7

WIPE OFF ANY PASTE INSIDE

4. Rule a line 41 cm from fold, level with fold. Cut off both ends together along line.
5. Open out the centre fold.

6. Paste a strip 10 cm wide at one end.
7. Place other end over pasted strip. Smooth out and leave under books to dry.

CREASE

8

25 CM

25 CM

CUT

9

10

FOLD MARKS

EACH SIDE IS 36 CM WIDE

11

8. When dry, fold in centre. Crease hard.
9. Rule 2 lines 25 cm apart, level with the top edge. Cut along the lines.

10. Unfold the 2 pieces. Mark the top and bottom edges to show where the folds are.
11. You now have the 2 cell covers.

The sticks and bracing pads

4 'FLY BEAD' STICKS 84 CM LONG

A

B

1·5 CM

12 CM

The 4 corner sticks

You will need to buy about 8 metres of fly beading (about 12 mm wide, 6 mm thick). Ask for 4 lengths, each 2 metres long. From each piece cut a stick 84 cm long.

Measure and mark the divisions A and B on one stick only – mark the other 3 sticks from this one. Draw the marks all around the sticks to make them visible from all sides.

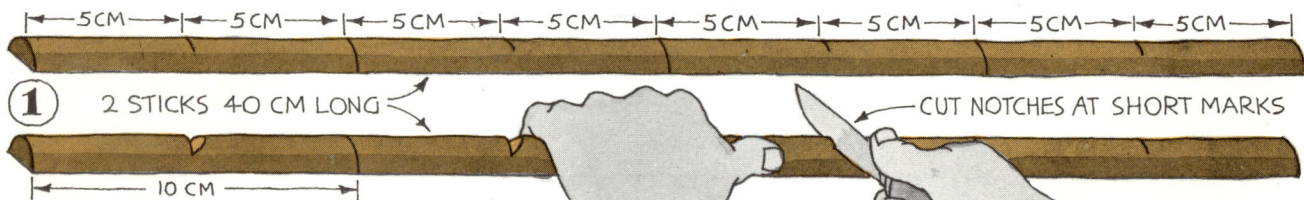

5CM · 5CM · 5CM · 5CM · 5CM · 5CM · 5CM · 5CM

① 2 STICKS 40 CM LONG

CUT NOTCHES AT SHORT MARKS

10 CM

The bracing pads

1. Cut 2 more sticks, each 40 cm long. Mark off each stick in 5 cm lengths. Make every **second** mark longer as shown.

2. Cut notches with your knife on each **short** mark. Cut each notch with one side vertical, and the other oblique. Make them about 3 mm deep and 4 mm wide at the top.

3. When the notches are cut, saw off the pieces at the long marks and you have 8 bracing pads.

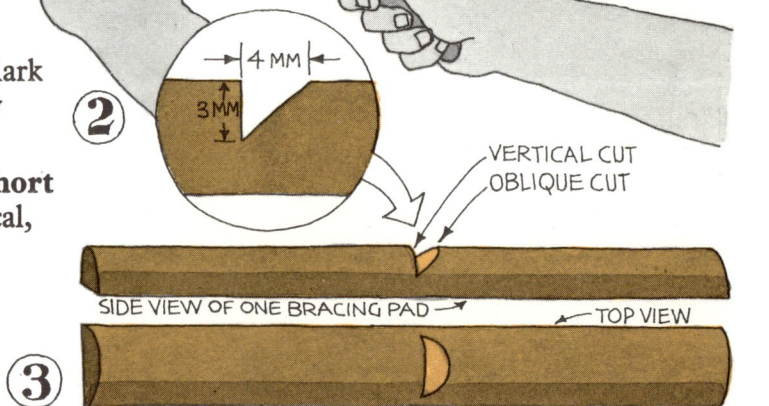

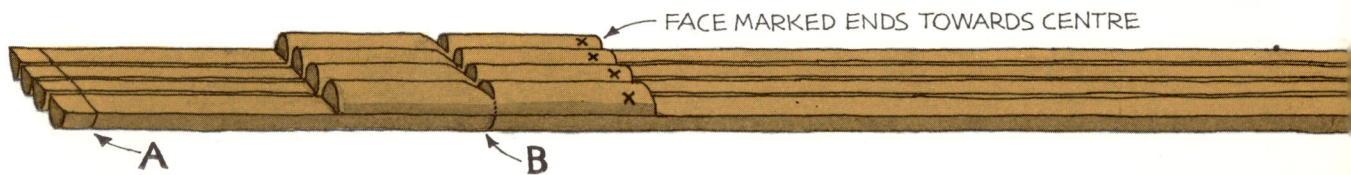

②

4 MM

3 MM

VERTICAL CUT
OBLIQUE CUT

SIDE VIEW OF ONE BRACING PAD →

TOP VIEW

③

FACE MARKED ENDS TOWARDS CENTRE

x
x
x
x

A

B

B → ← A

|← 12 CM →| |← 1·5 CM →|

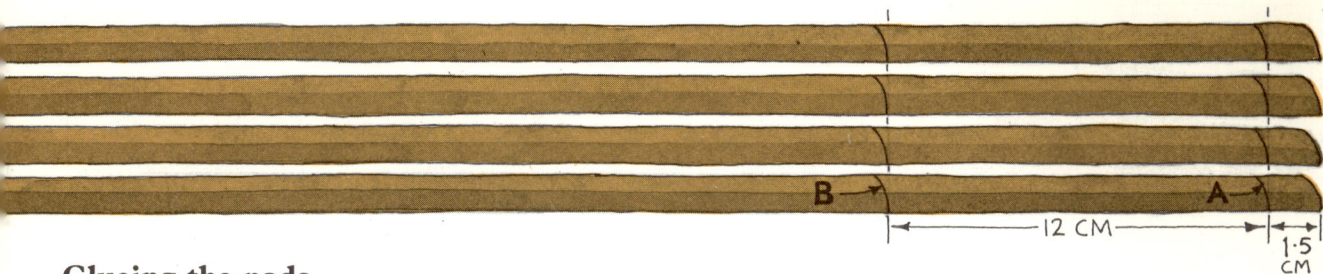

Glueing the pads

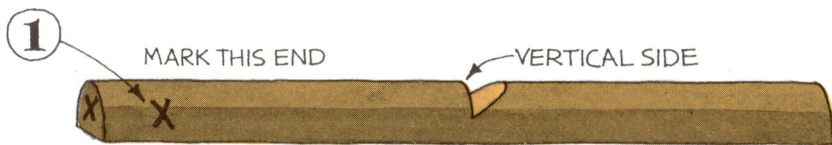

1 MARK THIS END — VERTICAL SIDE

1. On each pad, mark (with an X) the end closest to the vertical side of the notch.

2. Spread glue on the flat side of one pad.

3. Stick the glued pad to the flat side of one of the frame sticks. The notch should be directly above the mark B. Make sure the X marked end is nearest the centre of the stick as shown. Press the sticks together.

4. Glue on the other 7 pads, one at a time. Remember to place each pad with the X marked ends towards the centre of the stick. Lay the sticks somewhere flat to dry.

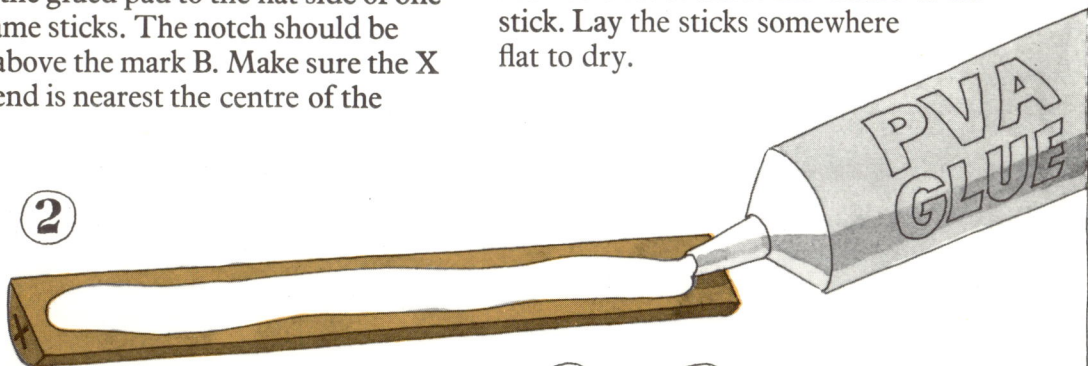

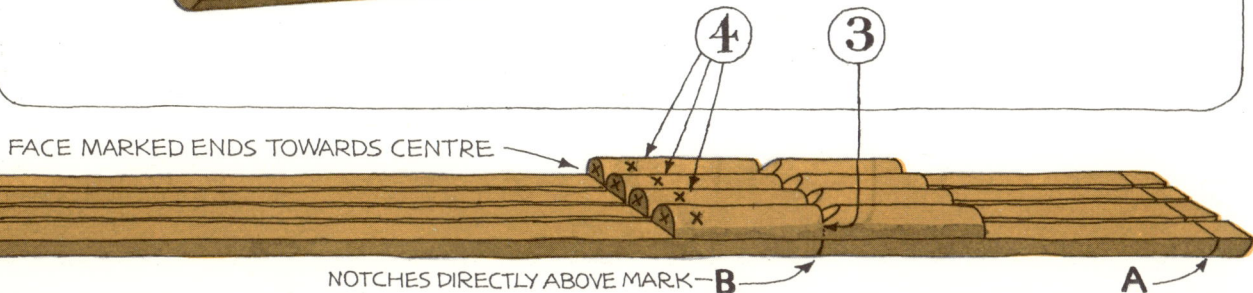

2

PVA GLUE

4 **3**

FACE MARKED ENDS TOWARDS CENTRE

NOTCHES DIRECTLY ABOVE MARK — B

A

Taping the covers to the sticks

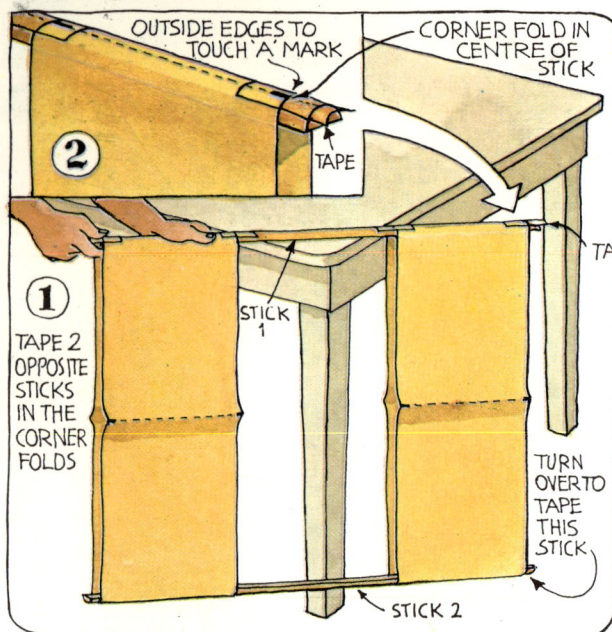

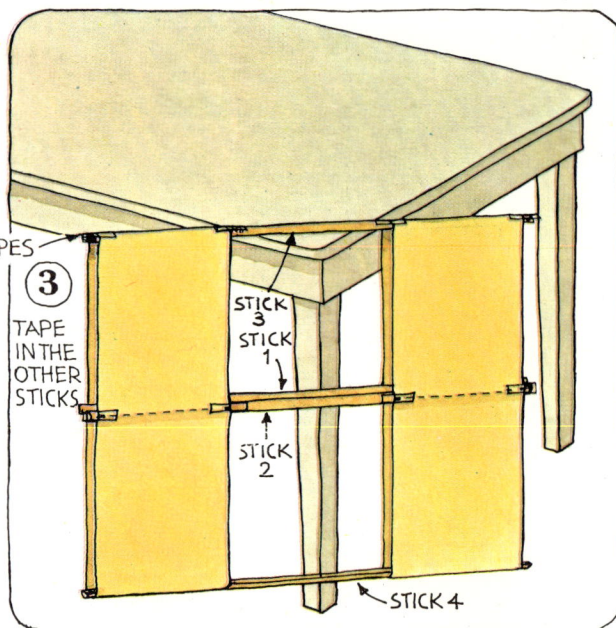

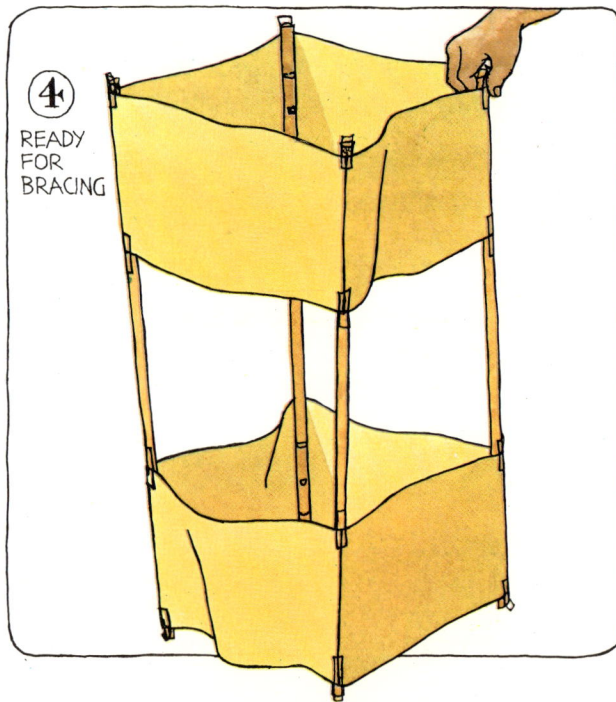

OUTSIDE EDGES TO TOUCH 'A' MARK

CORNER FOLD IN CENTRE OF STICK

② TAPE

① TAPE 2 OPPOSITE STICKS IN THE CORNER FOLDS

STICK 1

TURN OVER TO TAPE THIS STICK

STICK 2

③ TAPES

TAPE IN THE OTHER STICKS

STICK 3
STICK 1

STICK 2

STICK 4

④ READY FOR BRACING

1. Slip 2 sticks inside the 2 cell covers. Balance one stick on the corner of a table, with the other stick hanging underneath. Have the **curved side** of each stick resting in opposite corner folds of the cover.

2. Have the outside edge of each cover touching marks A on the top stick. The fold must run down the **centre** of the curved side of the stick. Tape them to the stick with 4 short pieces of tape as shown. Hold the sticks apart. Turn them over and hook the other stick over the table corner. Tape the covers to this stick in the same way.

3. Tape the covers to the other two sticks in the same way. Make sure the folds run down the centre of each stick.

4. Now the kite will stand up ready for bracing.

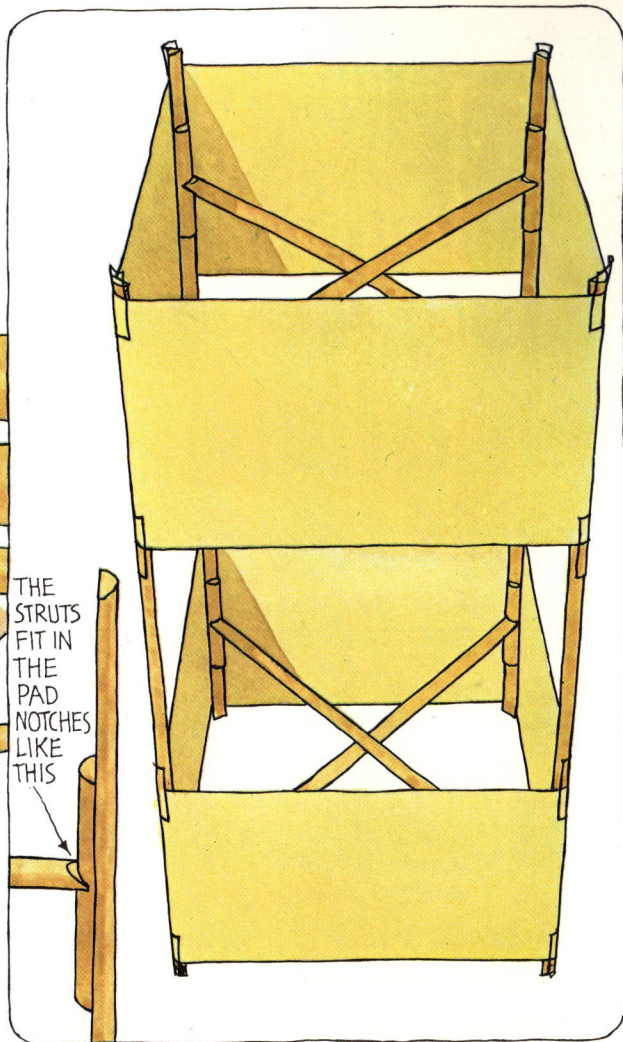

The bracing struts

①

As you can see in the drawing, each cell has 2 'struts' fitting tightly between the corner sticks. We brace one cell first before cutting struts for the second cell.

1. Start by cutting 2 sticks 49 cm long. Make sure the ends are cut square.

2. On one stick make a mark 5 mm back from one end, on the curved side.

←5MM→

②

③

1 MM FLAT END

THE STRUTS FIT IN THE PAD NOTCHES LIKE THIS

3. From this mark, cut the end of the stick across at an angle with your knife. Do not make the end sharp, leave a flat edge 1 mm wide as shown.

4. Cut the other 3 ends in the same way.

④

Bracing the cells

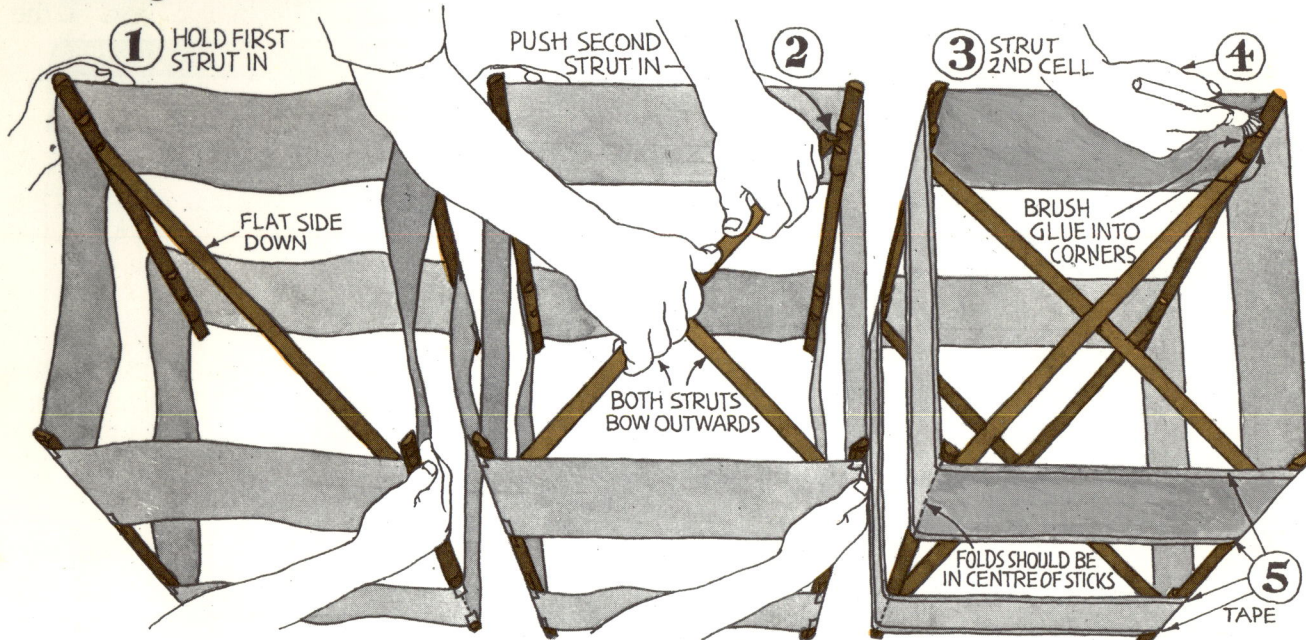

1 HOLD FIRST STRUT IN

FLAT SIDE DOWN

PUSH SECOND STRUT IN

2

BOTH STRUTS BOW OUTWARDS

3 STRUT 2ND CELL

4

BRUSH GLUE INTO CORNERS

FOLDS SHOULD BE IN CENTRE OF STICKS

5

TAPE

1. Stand the kite upright. Ask someone to hold it for you while you fit the struts. Place one strut (flat side down) inside the top cell. The ends fit in opposite pad notches.

2. While your helper holds the first strut in, place one end of the other strut in its notch. Push the other end into the opposite notch. This should be a tight fit. Ask a grown-up to help you if you are not strong enough to get it in. The struts must fit tightly so that the kite will be strong. They should bow **outwards** slightly (about 1 cm) when in place.

If you find the struts are very tight and need to bow up more than 2 cm, shorten them **both** by 1 mm. Loose-fitting struts can be made tighter by placing pieces of thin cardboard in the notches but it is better to cut 2 new sticks 1 mm longer.

3. When you have the first cell tightly strutted, cut two more sticks to the same size. Turn the kite over and strut the other cell.

4. The kite will now be quite rigid. Check that the corner marks are still in the centre of the stick. Adjust if necessary. Secure the covers by brushing plenty of P V A glue or paste into the inside cell corners. Brush it in hard, between the stick edges and the paper.

5. Remove the pieces of tape. Place long strips of tape all around the edges of the cells to help stop tearing. Press it down firmly.

There is no need to glue the struts into the notches. If they fit tightly they will come out only if the kite crashes heavily. Not glueing them means you can collapse the kite by taking them out. The collapsed kite can be made into a bundle, easy to carry about.

Flying the Sock Box

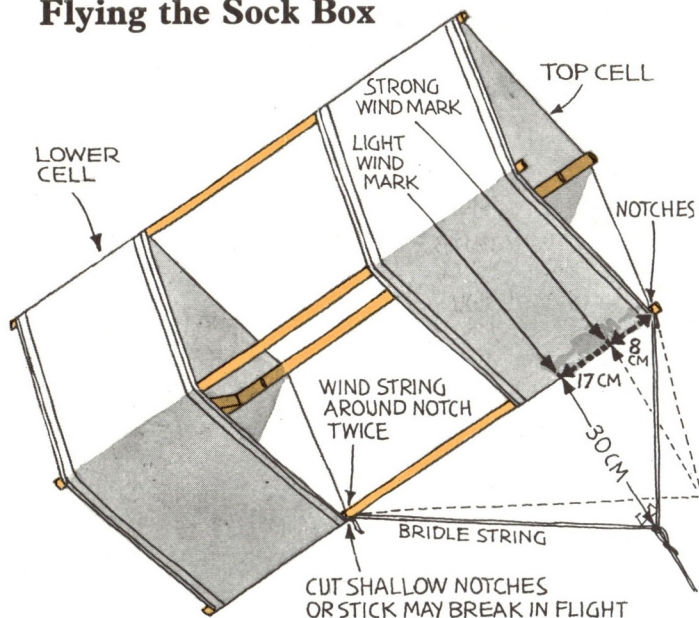

LOWER CELL

STRONG WIND MARK

LIGHT WIND MARK

TOP CELL

NOTCHES

8 CM

17 CM

30 CM

WIND STRING AROUND NOTCH TWICE

BRIDLE STRING

CUT SHALLOW NOTCHES OR STICK MAY BREAK IN FLIGHT

The bridle: Cut shallow notches 1 cm from the end of one corner stick. The cell at this end will be the top cell. Notch the same stick just above the lower cell. Cut a bridle string about 1 metre long. Tie one end to the top notch. Pass the other end twice through the flying line loop. Allow about 30 cm slack and tie the other end to the lower notch. The towing position will be opposite a point about 17 cm from the top edge of the top cell. For very strong winds move the towing position farther **up** the bridle string.

Flying: The cells of the Sock Box trap more wind than the Tadpole or Bat, so make sure your line is strong and in good condition. Choose a fairly windy day. If the wind falls below the strength necessary to support the kite, it will drop very quickly. If this happens you can prevent damage by running with the kite **into** the wind, slowing its descent.

To launch the kite stand it up on end as shown. A friend can hold it while you walk about 40 paces into the wind, letting out the line. Hold the reel tightly. Walk or run into the wind and the kite will rise into the sky. Experiment with the bridle until you find the best towing position. This is the only thing we can adjust on this kite. Good luck with it and may your Sock never turn inside out.

HURRY UP! I CAN'T HANG ABOUT HERE ALL DAY!

WIND DIRECTION

YOUR HELPER HOLDS IT UP LIKE THIS

HELP!!

WHOOSH!

DON'T WORRY— THIS COULDN'T HAPPEN.

ALWAYS GET YOUR KITE UP HIGH AS QUICKLY AS YOU CAN. THE HIGHER THEY ARE THE STEADIER THEY FLY.

Decoration

MAKE A 'MINUTEMAN' WITH THIS SHAPE

CRÊPE PAPER 'FEATHERS' OVERLAP

PASTE ON COLOURED PAPER STRIPS

PASTE

A brightly coloured kite looks most attractive, and offers a good opportunity to show off your skill at drawing and painting.

Kite shapes suggests many things. Let your imagination go. The best designs have simple shapes and bright colours. Thick black outlines make colours appear brighter. Use poster or acrylic colours.

Coloured streamers made from crêpe paper can be fixed to different parts of your kite. Be careful that they don't make the kite too heavy, or put it off balance.

Many flyers like to leave their kites without decoration. They just like flying them.

Whatever you decide, you will find kite making and flying a lot of fun.